Practical Guide to Rock Tunneling

T0179197

Practical Guide to Rock Tunneling

Dean Brox

Brox Consulting, Vancouver, Canada

CRC Press
Taylor & Francis Group
Boca Raton London New York

CRC Press is an imprint of the
Taylor & Francis Group, an **informa** business

A BALKEMA BOOK

Upper right cover page photo courtesy of Society for Mining, Metallurgy and Exploration

Published 2017 by CRCPress/Balkema
P.O. Box 447, 2300 AK Leiden, The Netherlands
e-mail: Pub.NL@taylorandfrancis.com
www.crcpress.com – www.taylorandfrancis.com

First issued in paperback 2021

© 2017 by Taylor & Francis Group, LLC
CRC Press/Balkema is an imprint of the Taylor & Francis Group, an informa business

Typeset by Integra Software Services Private Ltd

ISBN 13: 978-0-367-78218-4 (pbk)
ISBN 13: 978-1-138-62998-1 (hbk)

Visit the Taylor & Francis Web site at
http://www.taylorandfrancis.com

and the CRC Press Web site at
http://www.crcpress.com

Library of Congress Cataloging in Publication Data

Although all care is taken to ensure integrity and the quality of this publication and the information herein, no responsibility is assumed by the publishers nor the author for any damage to the property or persons as a result of operation or use of this publication and/or the information contained herein.

Contents

Foreword

"Practical Guide to Rock Tunneling" is exactly that, a useful tool and guide for practitioners in the tunnel business. This book provides insight into nearly every subject and for all steps of a rock tunnel project from planning through construction and operations and inspection. This book is a valuable reference for owners/clients, as it includes discussion of subjects such as: project planning, costing, scheduling, and risk assessment, never covered in classic technical books. Excellent photographs and graphics are contained throughout the book, which clearly illustrate the author's message from practical experience on the multitude of subjects.

This book provides guidance on the current up-to-date state-of-the-practice in rock tunneling discussing innovations of the last decade from televiewer logging of bore holes, LiDAR aerial surveys for geological fault mapping, proper photography of rock core, to the modern use of fiber reinforcement in pre-cast concrete segments. New technical evaluation and graphical methods are presented for overstressing and squeezing of deep tunnels as part of constructability assessments. A good discussion, illustrated with photographs, of the importance of rock durability is provided. The acceptability of unlined tunnels for water conveyance and criteria to consider are discussed.

This book is based on 30 years of practical experience of investigation, design, construction, and operation of rock tunnels coupled with a thorough review of the literature. "Practical Guide to Rock Tunneling" should be part of all tunnel practitioner's libraries; this includes staff engineers, geologists, project managers, and owners.

Don W. Deere, P.E.
Denver, Colorado

About the author

Dean Brox has over 30 years of experience as a tunneling practitioner with more than 1200 km of major tunneling and infrastructure projects around the world. Dean graduated in 1985 from the Geological Engineering Program of the University of British Columbia in Vancouver, Canada, in 1985 and obtained his Master in Science in Engineering Rock Mechanics with Distinction from Imperial College, London, in 1990. Dean started his career in underground mining rock mechanics in South Africa before moving into tunneling for civil engineering infrastructure projects around the world including the Lesotho Highlands Water Project, the Hong Kong Airport Core Projects, and the Gotthard Base Rail Tunnel in Switzerland. Dean has extensive experience with tunnels for hydropower projects and his areas of particular interest include high speed drill and blast excavation of long tunnels, long and deep TBM excavated tunnels, overstressing of deep or weak rock tunnels, asset performance of hydropower tunnels and TBM applicability assessments for mining projects. Dean practices as an independent consulting engineer and lives in Vancouver, Canada.

Acknowledgements

As with many tunnel engineers, I happened to fall into this profession by chance after starting my career working as a rock mechanics engineer in the deep level gold mines in South Africa, and then being part of the design team for Phase 1A of the Lesotho Highlands Water Project that included more than 80 kilometers of tunnels.

In 2016, after 30 years of practice, I can now look back and consider myself to have been very fortunate to have had world-class mentors during my early years in the industry who taught me the fundamentals of tunneling and the associated intricacies of the practice. I wish to acknowledge: Alan Earl, T.R. (Dick) Stacey, the late Mike Dewitt, Jim Richards, John Sharp, Ian McFeat-Smith, and the late Heinz Hagedorn. I also wish to acknowledge additional professionals who I have continued to learn from during my recent years in the tunneling profession including Don Deere, Evert Hoek, Ray Benson, Andrew Merritt, Einar Broch, Allan Moss, and Laurie Richards. Finally, I wish to thank all the colleagues whom I have enjoyed working and sharing experiences with during my career to date from South Africa, Hong Kong, Switzerland, Chile, the USA, India, and Canada.

Dean Brox, Vancouver, Canada

Dedication

This **Practical Guide to Rock Tunneling** is dedicated to the memories of my parents, Gordon and June Brox who inspired me to explore the world in this profession of tunneling and share my experiences and lessons learned and provide a contribution to help educate others for the future.

Disclaimer

No responsibility is assumed by the Author or Publisher for any injury and/or damage to persons or property as a matter of products liability, negligence or otherwise, or from any use or operation of any methods, products, instructions or ideas contained in the material herein.

Figures and Tables

Figures

Tables

Chapter I

Introduction

The majority of tunnels that are planned and constructed for various forms of infrastructure requirements around world are located in urban areas and are sited at shallow depths in soils or overburden above bedrock. An increasing number of tunnels are however now being planned, designed, and constructed around the world at greater depths within bedrock, particularly for hydropower and mining projects, but also to a lesser extent for civil infrastructure. Over the past decade several of these such tunnels have experienced some serious problems during construction due to limited understanding of the key challenges associated with their design, construction, and operation. Tunnels planned to be constructed in bedrock are associated with their own unique design requirements and challenges due to the commonly recognized high variability of the geological conditions, its behaviour, and its influence on the design, construction and operation of tunnels.

A unique difference between tunnels in rock versus most of those tunnels located in soils is the possibility to allow the rock to form as the final internal surfaces of the tunnel and be exposed and not be fully lined by shotcrete or concrete which commonly presents economic benefits for a project. The acceptability of partially unlined or non-lined tunnels in rock, especially those for water conveyance, represents one of greatest challenges for tunneling practitioners as getting it wrong either during construction, but more during operations, can have significant cost impacts associated with repairs and loss of service. Several tunnel projects have benefitted from allowing the tunnel to be partially unlined where the quality of the rock conditions has been fully evaluated and confirmed to be durable and therefore acceptable for the intended service of long term operations.

It is important for tunneling practitioners to recognize that the planning, design, construction and operations of tunnels is a global-experienced based profession. The purpose of publishing the **"Practical Guide to Rock Tunneling"** is to pass on some relevant global and practical experience and lessons learned to the next generation of global tunneling practitioners, educators, and decision makers involved with rock tunnel projects.

This **"Practical Guide to Rock Tunneling"** is intended as a practical road map for the tunneling practitioner for the design, construction, and operations of tunnels in rock. As a noted handbook it provides recommendations for good practice for the industry, and in some cases, guidelines that are considered to represent good industry practice, including lessons learned from past projects. It is not intended to provide an exhaustive account of detailed information on each sub-topic of rock tunneling that has already been published and rather offers key references that can be further searched if required by the reader. While this **"Practical Guide to Rock Tunneling"** focusses on tunnels in rock, many of the following subjects are applicable to other types of underground excavations including shafts, chambers, and caverns.

The **"Practical Guide to Rock Tunneling"** is aimed at undergraduate and postgraduate students, young professionals starting in the tunneling industry, as well as individuals who do not have an extensive background in tunneling but who are responsible to make decisions for planned tunnel projects. The **"Practical Guide to Rock Tunneling"** takes the reader through all the critical steps of the design and construction for rock tunnels starting from the execution stages of a tunnel project, geotechnical site investigations, rock characterization, design, evaluation of risks, construction considerations, and through to construction supervision and post-construction inspections for safe future operations. The **"Practical Guide to Rock Tunneling"** also presents suggestions and recommendations for tunneling practitioners on special topics of laboratory testing, durability of rock and acceptance for unlined water conveyance tunnels, overstressing or deep and long tunnels, risk-based evaluation of excavation methods, contract strategies, and post-construction inspections. Key considerations and lessons learned from selected case projects are presented based on the author's extensive international experience of over 30 years and 1000 km of tunneling for civil, hydropower, and mining infrastructure, including some of the most recognized projects around the world to date.

Tunneling is a practice that will continue to evolve with the development of new procedures and codes, methods of analyses for design, and technologies that will allow for the construction of tunnels in the future using different approaches than in the past. Tunneling practitioners should take the opportunity to attend tunneling conferences and seminars as well as courses to increase their knowledge of all of the aspects of tunneling. Tunneling practitioners should also read and review journals, conference proceedings and notes from seminars and courses to keep up to date on current advances and developments in tunneling technology. Various websites including those of most journals now provide an incredible wealth of very useful information on new tunneling projects, as well as the status and progress of current projects, including challenges during construction, and solutions implemented. The following websites are available:

- TunnelTalk
- Tunnel Business Magazine
- Tunneling Journal
- Tunnels and Tunneling International
- Tunnel Builder
- Tunnel (German and English)
- AFTES (French and English)

Finally, senior tunneling practitioners should make their best attempts to devote some of their time for the mentoring of young tunneling practitioners in order to transfer valuable experience from past projects to the next generation of tunneling practitioners.

Chapter 2

Functional uses of rock tunnels

2.1 General

Tunnels in rock are being planned, designed and constructed for an increasing variety of plausible and environmentally accepted solutions for infrastructure requirements around the world. As space on the surface becomes increasingly limited and congested, particularly in urban areas, the use of underground space with tunnels can be expected to be identified as a viable solution for the transport of people, materials for everyday living.

In addition, existing and dormant tunnels are increasing being evaluated for alternative uses in society to take advantage of their pre-existing status and the application of well proven renovation and rehabilitation techniques for their transformation.

2.2 Functional uses

Tunnels designed and constructed in rock have historically been used and continue to be used for a wide variety of common infrastructure requirements including the following:

- Access;
- Bicycle;
- Conveyor;
- Combined Sewage/Stormwater Overflow (CSO);
- Drainage;
- Drinking Water;
- Exploration;
- Hydropower;
- Pedestrian;

- Pipelines (oil and gas);
- Rail (Light Transit and Heavy Freight);
- Sewage;
- Strategic storage (military);
- Traffic;
- Utilities (electrical cables, fibre optics);
- Water Diversion;
- Water Supply (Irrigation)
- Wine caves/storage, and;
- Ventilation.

Tunnels in rock have also been designed and constructed for very specialized purposes including nuclear particle physics research as at CERN (European Organization for Nuclear Research), Switzerland and Stanford University in the United States. Several tunnels in rock (and other types of excavations) have also been designed and constructed as part of ongoing research of the storage of nuclear waste. Many tunnels in rock have also been designed and constructed for multiple purposes including mine access and conveyor, access and ventilation/utilities, flood control and traffic such as the Stormwater Management and Road Tunnel (SMART) in Kuala Lumpur, Malaysia

(SMART Tunnel), and hydropower generation and irrigation. Figure 2.1 illustrates a historical rail tunnel built in the early 1900s that has been converted into a walking trail within a nature reserve in western Canada.

A recent resurgence in the demand for energy and in particular renewable energy has resulted in the design and construction of several hydropower tunnels around the world. In addition, many urban areas continue to face transportation and stormwater control challenges and therefore an increasing amount of metro/subway and CSO tunnels are being built in many large cities. Furthermore, the use of underground space, and in particular, the conversion of historical and dormant tunnels is being recognized for new transportation solutions such as bicycle and pedestrian tunnels in many cities.

Figure 2.1 Historical rail tunnel.

Chapter 3

Tunnel project execution

3.1 General

Tunnel projects are required to be executed in an appropriately planned sequence if they are expected to be successful and not associated with major delays. Clients and the developers of tunnel must recognize the importance of pre-planning and that key decisions are required to be made in order to allow key stages of a tunnel project to be advanced to the next stage. One of the first decisions required to be made is the method of delivery, that is, the procurement approach for the project. A project execution plan including a total project schedule is recommended to be prepared by the developer or client upon conceptual realization for a new rock tunnel to address all of the key requirements for a successful completion.

3.2 Project delivery method

Prior to the start of a tunnel project the client is required to select the method of project delivery for the project. The main two types or methods of project delivery are:

- Design-Bid-Build, and;
- Design-Build.

Design-bid-build (DBB) is the traditional method of project delivery whereby the client typically engages a tunnel consultant to prepare a final design that is bid upon by pre-qualified tunnel constructors and is awarded to a preferred tunnel constructor based on some form of evaluation criteria. The DBB approach generally require a longer total period of time for the completion of the project but has the advantage of the complete control of the design requirements by the client through each stage of design and construction. The DBB approach should be adopted as the project delivery method for all hydraulic tunnels owing to the typical complexity of the design associated with hydraulic tunnels and their associated works.

Design-build is essentially a fast-track approach of project delivery that typically requires a shorter period of time for project completion whereby the tunnel consultant engaged by the client only prepares a reference design in conjunction with a design criteria of the functional requirements which does not represent a final or detailed design. The reference design is then bid upon by pre-qualified DB teams comprising tunnel constructors and their designated designers and is awarded to a DB team based on evaluation criteria commonly including best price and best schedule. The DB approach has the advantage of attracting innovation into the design process by the DB team but the disadvantage of limited control over the final design and construction. The DB approach has been successfully used for several types of tunnel projects. The DB approach is not considered to be ideal for complex projects where the final design details may change significantly during construction or are critical for operations such as for hydropower projects. Various aspects of the design-build project delivery approach are presented in Brierley *et al.* (2010).

3.3 Execution stages

It is important for all parties involved to understand the common stages of the execution for a tunnel project. Although there are strong differences between the design requirements between the civil and mining engineering industries, the overall planning and execution stages for a tunnel project are similar.

The typical stages of execution for a rock tunnel project include the following:

- Conceptual design
- Options trade-off/order of magnitude
- Pre-feasibility;
- Preliminary design/feasibility;
- Basic and Detailed Design;
- Construction;
- Commissioning, and;
- Operations.

For some complicated projects it is not unusual that multiple attempts of conceptual and even preliminary design stages are completed as part of an overall project optimization or in the case of changes in environmental or local governmental regulations. For a Design-Build project delivery method it is common that a reference design is prepared for bidding after the conceptual, pre-feasibility or feasibility stage of the project, subject to the project schedule.

3.4 Pre-planning by client

Most tunnel projects are initiated with a conceptual design or project definition including definition of a corridor or area within which the tunnel is expected to be located. An environmental study is commonly commenced and performed by an independent consultant under the direction of the client shortly after or concurrently with project definition. The conceptual design study or project definition should include the identification of all viable tunnel alignments within the tunnel corridor that are constructible using currently available technology and also meet the operational requirements for the intended function. The conceptual study should include a comprehensive technical comparative evaluation of the tunnel alignments based on key criteria important for the project such as cost, schedule and risks, and should include risk ratings and ranking of the alignments in terms of constructability in order to create a shortlist of the tunnel alignments with a preferred alignment to focus any planned geotechnical investigations on a limited number of alignments. Technical pre-feasibility and feasibility stages can then proceed and be confirmed subject to the initial findings of geotechnical investigations and/or local experience. Feasibility is a critical step and gate decision for project development and serves as a basis for investment decisions and as a bankable document for project financing. Feasibility studies should be performed to confirm that the project is technically and commercially sound and ensure that all of the risks have been identified and mitigation established.

Some clients have internal engineering departments and groups with experienced resources to perform the conceptual engineering or option trade-off studies or at least to identify the corridor for the tunnel project, which in some cases is defined due to external reasons such as environmental restrictions. If a client does not have internal resources for the completion of conceptual engineering then an independent consultant should be engaged and managed by a designated project manager of the client to direct and review the work of the consultant. The client should fully recognize the importance of having a designated project engineer or project manager to oversee the services to be provided by the independent consultant on a regular basis that includes frequent meetings and presentations of the status of the project engineering.

The selection of an experienced independent consultant by the client should be based on a thorough review of information submitted from an expression of interest and an interview process, and possibly further from a review of information submitted from a request for proposals. The client should review and confirm the quality of services provided by independent consultants during past projects based on client references.

For most tunnel projects it is in the interest of the client and for the success of the project to engage the independent consultant for all stages of project engineering as well as construction oversight and supervision such that the responsibility of the design, and any modifications during construction, is fully maintained by the original project designer and Engineer of Record.

3.5 Project engineering by consultants

The level of effort that is expected by the client to be completed by the independent consultant should be thoroughly presented and discussed at the start of the engineering services. The level of effort required to be completed as part project engineering for a tunnel project does not vary significantly with the size of the tunnel project. The engineering services should be executed in conjunction with a design engineering schedule with key milestones and deliverables. The typical project engineering services to be provided by the independent consultant includes the following:

- Project Management;
- Design Criteria and Basis;
- Technical Memoranda;
- Risk Registers;
- Geotechnical Data Report;
- Preliminary design – 30%;
- Preliminary design – 60%;
- Preliminary design – 90%;
- Final design – 100%;
- Technical Specifications;
- Tender Drawings;

- Geotechnical Baseline Report, and;
- Form and Conditions of Contract and Bill of Quantities.

The preliminary and final design deliverables of each milestone should include design drawings, and opinion of probable construction costs and construction schedules. Technical memoranda should be required to be completed for each key design subject and component of the project with updating of the design criteria and basis as the design is advanced and completed. For some tunnel projects additional specialist sub-consultants commonly form part of the design team for geotechnical, environmental, hydraulics and possibly mechanical and electrical services. The client should perform independent reviews at each stage of design.

It is also common practice during the early stages of a design process that the tunnel consultant will be expected to provide key input to the environmental study undertaken by a specialist consultant to confirm probable construction sites for access, laydowns and spoil disposal.

3.6 Engineering effort and deliverables during execution

The engineering effort and deliverables that should be completed at each stage of a tunnel project should be clearly defined by the client at the start of the services and thoroughly reviewed on a regular basis during project execution.

The typical definition of the engineering effort and key deliverables necessary to be completed for each stage of the project execution is presented in Table 3.1 below.

Table 3.1 Engineering effort and key deliverables.

	Conceptual	Pre-feasibility	Feasibility	Detailed
Design and Engineering Effort	Simple designs showing concepts, organization of layouts etc. Identify significant items which may involve complex technology, advanced design or research. Identify constraints and opportunities for cooperating with government and international organization on transport, power and water infrastructure. Conceptual study report with drawings/ figures of layouts and typical designs. Initial risk register, typically 10%	Preliminary site layouts with all major components to scale but with limited detail. Identify transport routes (including rail/port etc.). Designs developed in response to processing plant designs, power and water requirements, and transport routes. Pre-feasibility report with drawings/layouts typically 30%	Preliminary layouts with additional details of building designs for major items (i.e. rail, roads, support facilities, pipelines, power, ports etc.). Feasibility Report with drawings/ layouts presenting details, preliminary technical specifications, preliminary bill of quantities, preliminary geotechnical baseline report, typically 60%	Final layouts with detailed building designs for major items (i.e. rail, roads, support facilities, pipelines, power, ports etc.). Detailed Design Report with final drawings, final technical specifications, and bill of quantities, Form of Contract, pre-qualification documents, 100%

3.7 Functional requirements for design build

If the client prefers to adopt a design-build approach for the tunnel project then the functional requirements for the proposed tunnel need to be prepared and be inconclusive of all key operational requirements including design standards for design-build teams to prepare proposals.

If the client does not have technical resources with appropriate experience in tunnel design and construction and preparation of functional requirements then a tunnel consultant should be engaged to provide this service and also be utilized to review and evaluate design-build proposals.

The design-build approach also importantly requires a well-experienced tunnel design manager as part of the designer's team to frequently liaise with the tunnel constructor t during the development of the final design and to present the final design to the client for acceptance.

3.8 Early contractor engagement and involvement

In some cases it may be appropriate to engage a well experienced tunnel constructor during the early stages of the planning and design process. The experience of a tunnel constructor during these early stages can be invaluable for a tunnel project to provide insight into perceived risks, constructability issues, and as well as can be required to prepare an accurate cost estimate and construction schedule. It is common that tunnel constructors only be engaged during the early stages of a design as long as they are not excluded or disqualified later from bidding on the project. It is therefore important to decide at what stage the tunnel constructor should no longer be engaged.

The project delivery method may also adopt the full involvement of a preferred tunnel constructor selected based on qualifications from an early stage submitted proposal and be an integral part of the design team, together with the tunnel consultant and client, to develop and optimize the design for construction. This approach is referred to as Early Contractor Involvement (ECI). ECI represents a collaborative effort between the client, designer, and the selected constructor and is best implemented as early as possible within the project schedule to get the most benefits. The main benefits of ECI are improved constructability, cost estimation, and risk management as well as better planning for construction (Sødal et al., 2014). Distinct disadvantages of ECI are the lack of competition for pricing of the project and possible innovation, in comparison to the Design-Build approach.

The tunnel constructor will be required to provide prices for the construction of the project, thus serving to confirm the anticipated cost to an accurate degree at an early stage for the client. If the client does not accept the prices prepared by the tunnel constructor then there can be the opportunity to invite alternative bidders to compete for the award of the construction of the project. This method of procurement is ideally suited for fast-track projects where cost certainty is vital early on.

The ECI approach may also include the selection of multiple shortlisted tunnel constructors as a means to maintain competition with the requirement for each shortlisted team to develop the design-build reference design to completion with pricing. However, this approach will require greater effort by the client's team to evaluate each of the proposals and often includes the payment of a stipend to the non-selected tunnel constructors, so will incur higher costs to the client.

3.9 Constructability reviews

Many project teams have trouble determining issues that a project will face before experiencing them. Constructability reviews should be undertaken at each stage of design to evaluate the overall feasibility of the entire construction approach for the project including:

- Site layout and access;
- Practical and efficient designs;
- Construction staging and sequencing;
- Critical path activities;
- Temporary works requirements;
- Method of excavation;
- Equipment to be used;
- Identification of construction risks;
- Mitigation measures to be applied;
- Alternative means and methods;
- Potential innovation;
- Achievable productivities;
- Inconsistencies between plans and specifications;
- Material compatibility, and;
- Impacts on project cost, schedule, functionality, and suitability.

Constructability reviews may be performed by the design team but there are benefits to undertaking the reviews by independent external experts or the designated construction management team to introduce fresh eyes to help ensure that the project design drawings and specifications are efficient and readily buildable as per schedule and budget.

Constructability reviews may include the reviews of completed geotechnical investigations, construction cost estimates, and project schedules.

3.10 Independent TBM risk assessment

Based on current industry practice there appears to be recurring instances where TBM suppliers are not provided all of, or a concise version of, the relevant geotechnical information in order to advise their clients, who are typically the bidding Tunnel Contractors but may also be Project Owners who wish to consider pre-purchasing, of the most appropriate type of TBM and/or important components to include, that meets the project requirements and specifications, or may be in the interest of the project for risk planning and management.

This is recognized as a "gap" in the current industry practice that should be appropriately addressed in order to prevent an incomplete understanding of the expected tunneling conditions and overall tunneling risks, which TBM suppliers should be knowledgeable about, or well understand, in order to correctly advise their clients accordingly, and include all relevant provisions in the selection and design of the TBM in order to attempt to reduce construction risks.

The proposed method to address the "gap" is to create a "TBM Risk Assessment" report that should be completed on behalf of the Project owner/developer, and provided to all bidders at the time of bid (along with all other respective contact documents) as well as to all prospective TBM suppliers (or those specifically pre-qualified).

The TBM Risk Assessment report should be compiled by a recognized industry expert in TBM tunnel design and construction and not just any practicing geotechnical or geological consultant who may not be fully experienced in the use of TBMs and TBM tunnel construction risks.

The TBM Risk Assessment report, may, upon identification and review of all tunnel construction risks, propose a particular type of TBM, but should not be intended, in any way, propose a particular TBM manufacturer, or even details such as type of TBM, and rather leave that decision up to the selected Tunnel Contractor, unless there is a consistent historical practice of success of a particular type of TBM for the anticipated geological conditions.

The TBM Risk Assessment report will serve a good and strong purpose for Owners/ Developers, Tunnel Contractors, as well as TBM manufacturers in the industry to reduce the burden placed upon Tunnel Contractors during bidding of having to typically quickly review and summarize the relevant geotechnical information, which for some projects, may be voluminous.

Figure 3.1 Double shield TBM for the 12 km Vishnugad Pipalkoti Hydropower Project.

The TBM risk Assessment report is not envisaged to form part of the contract documents of any project like a Geotechnical Baseline Report, but portions of the report may be contained in any project risk register, that is becoming encouraged and more commonly used in the industry and to form part of the contract documents. Figure 3.1 presents the double-shield TBM that was uniquely designed by the selected tunnel constructor and incorporates risk reducing features and components based on lessons learned from previous similar projects for the construction of the 12 km pressure tunnel as part of the Vishnugad Pipalkoti Hydropower Project in India.

Chapter 4

Site investigations

4.1 General

One of the most frequent questions posed by clients is "How many drillholes are necessary as part of a site investigation program?" Clients should appreciate and understand that there is no simple and automatic answer to the question and tunneling practitioners should not attempt to provide any such answer if ever posed. Rather, a comprehensive approach and sequence should be adopted for the planning and execution of a site investigation program for a tunnel project that is based on a comprehensive evaluation of the perceived geological conditions, complexities, and uncertainties.

The level of effort of site investigations should be sufficient to define with confidence the "average" or "typical" subsurface conditions in terms of main rock types and their geotechnical properties. This level of investigation is necessary in order to be able to compile a representative geotechnical baseline report (GBR) such that significant claims do not occur during construction do to encountering differing site conditions that are more adverse than the presented baseline conditions.

As a general guideline, Parker (2004) suggests that it is prudent to consider that 2 to 4% of the total capital cost should be allocated for a site investigation budget. For a proposed rock tunnel project with an expected total cost of $100 Million Dollars, this amounts to about $2 to 4 Million Dollars. For a project site with complex geological conditions higher budget amounts should be allocated extending up to 6% of the expected total capital costs. For remote project sites it may be necessary to utilize helicopters for access which will result in significantly higher costs for the mobilization of drilling equipment, and transport of site supervision staff, and in some case for the supply of water. In general, all main structures or components of a tunnel project including portals, shafts, caverns, and key identified targets of possible high risk such as geological contacts, fault/shear zones, and inferred infilled valleys or paleochannels should be investigated with boreholes.

There are however increasing expectations in the underground industry for a prudent level of site investigations to be performed. For example, owners are at risk of only receiving a limited number or higher than expected bids if limited geotechnical data is available and the practice of some tunneling constructors is to withdraw from submitting a bid when there is limited data. In addition, lending agencies as well as insurers now require appropriate site information before their involvement in any particular project may be confirmed.

4.2 Potential consequences of limited site investigations

Figure 4.1 presents a relationship between the percentages of cost overruns to the total project cost versus the ratio of the total length of drilling to tunnel length for underground hydropower project sponsored by the Work Bank (Hoek & Palmieri, 1998). The relationship clearly shows a negative exponential relationship and can be considered as a useful tool for the planning and budgeting of a site investigation program whereby, for example, in order to reduce a possible cost overrun of 10%, the total amount of drilling in meters that is required should be equivalent to about 73% of the total tunnel length. Similarly, in order to limit a possible cost overrun of 5%, the total amount of drilling that is required is equivalent to about 100% of the total tunnel length.

A serious consequence of the completion of limited site investigations includes the submission of a limited number and commonly higher than expected construction bid prices. Tunneling constructors are increasing becoming more risk adverse due to the historical occurrence of unexpected adverse subsurface conditions being encountered during construction when limited site investigations were completed. In some cases, significantly higher than expected bid prices have been submitted and resulted in major delays to the start of the project since additional funding had to be secured by the client.

A further consequence of the completion of limited site investigations is the concern and possible rejection of, or enhanced cost to the client for insurance coverage for the project by the international insurance industry. The International Tunnel Insurance Group (ITIG, 2006) have established requirements for good industry practice for tunneling projects to address construction risks that includes the completion of an appropriate amount of site investigation prior to construction.

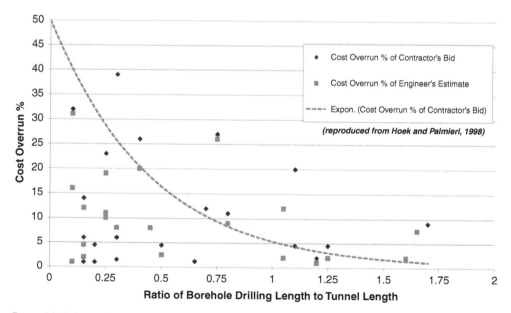

Figure 4.1 Relationship between drilling ratio and possible cost overrun.

4.3 Review of existing information and previous experience

Before proceeding with site investigations for a proposed rock tunnel project it is prudent to research and review any existing relevant information on the subsurface conditions as well as environmental conditions at the site from historical, recent and current projects. It is also very useful to visit and inspect the site assuming access is possible to view existing natural bedrock outcrops and possible rock cuts associated with roads and foundations. Environmental baseline study reports if available, can be a useful source of initial information about a particular site that should also be reviewed.

Relevant existing information may be available from historical underground projects in the project area through conference papers and public reports. Another possible source of relevant information may be found from university research projects (such as dissertations/thesis) and/or government related research projects as well as national geological mapping authorities. Finally, for remote project sites, another possible source of relevant information may be found from past mining exploration programs and the associated reports that are typically required to be registered with the local and/ or national government mining authority. In some countries including Canada and Chile there exists online (geographic information system) GIS based mapping services that allows the creation of coarse geological maps presenting the main rock units/types and main geological fault zones.

4.4 Planning and budgeting for site investigations

The undertaking of a site investigation program for a rock tunnel project should not be assumed to be a minor effort but rather requires comprehensive planning and organization which is fully justified given the expense of most programs. Clients of proposed rock tunnel projects are also well advised to dedicate their own internal staff resources to manage their designated consultants and be fully available during the execution of the field works since decisions commonly have to be made concerning program costs and often environmental and community issues.

Firstly, it is important to recognize that a site investigation program should be planned to be flexible in terms of decision making and changes allowed to be made during the program to the type and quantity of work. It is also important that the site investigation program is carried out in a prudent sequence of execution and, likely, in multiple phases. The sequence of the work is important in order to allow key decisions, with possible changes, to be made after discovery of initial information in order to minimize unnecessary costs and time delays to the overall program. A phased approach to a site investigation program is also important since it is common that not all of the required information is gathered as part of first phase "Phase 1" of the overall program and a possible Phase 2 and even Phase 3 programs may be necessary to target "data gaps" along the proposed tunnel corridor.

The typical good industry practice for the sequence a site investigation program is as follows:

- Desk Study (with base map);
- Topographic survey (Orthographic photographs and LiDAR);

- Geological Field Mapping;
- Geophysical Surveys;
- Drilling (including logging and photographs);
- In Situ Testing;
- Sampling;
- Laboratory Testing, and;
- Field Instrumentation and Monitoring.

For most rock tunnel projects it is assumed that the client has either undertaken their own internal conceptual study design or has engaged a tunnel design consultant for the first stage of the project. For the execution of a site investigation program it is necessary to engage an independent and well experienced geological/geotechnical consultant in conjunction with a tunnel design consultant to be responsible to manage and supervise all of the field, in situ, and laboratory testing works. The scope of the services for the geological/geotechnical consultant should include the preparation of technical specifications for the site investigation works as well as compilation of a factual data report of all information together with a preliminary interpretative report of the findings of the site investigation program. These deliverables are typically then handed over to the designated preliminary or final tunnel design consultant, and after review of the available information, it may be determined that additional site investigations are required as part of design modifications or changes as part of design development for the project.

An important planning aspect for a site investigation program is the pre-qualification and availability of drilling and in situ testing contractors. It is common in the industry that there are only a limited number of available and well experienced drilling and in situ testing contractors for any given tunnel project and it is important to plan ahead and contact such companies to gain their interest for the submission of their qualifications. In some cases it may also be important to engage specialist in situ testing contractors and also require their pre-qualification and confirm their availability. After shortlisting of acceptable drilling and in situ testing contractors it is useful to provide a preliminary bill of quantities in order to request a preliminary quotation to confirm the expected costs of the program. The next important step is to finalize the preferred drilling and in situ testing contractor in order to pre-book or reserve their team of workers and equipment for the project in case they are also competing for other projects at the same time which is typical. Comprehensive meetings should be held between the designated geological/geotechnical and tunnel design consultants together and the preferred drilling and in situ testing contractor to confirm the acceptability of the proposed field equipment and materials, including spare parts, and procedures to be adopted well prior to the start of the field work.

Finally, a comprehensive communication and responsibility protocol should be established between the client, the drilling and in situ testing contractor, and the consultants to be used during the execution of the site investigation program.

Of particular interest for most clients is the anticipated cost prior to execution for the completion of a site investigation program. The total costs for a site investigation program (per Phase) should include the costs of the investigation works at the site (geophysics, drilling/testing etc.), and supervision of site works by the designated geological/geotechnical consultant. Within the cost estimate for the proposed site investigation program a prudent contingency should be included for possible additional or delayed activities such as extra drilling lengths or additional holes and testing. In conjunction with the cost estimate for the program a detailed execution schedule should also be prepared for planning and monitoring purposes. The program schedule will serve to aid in decisions of optimizing the movement and use of all the field equipment in order to minimize standby time and charges to the project.

A useful technical reference for the evaluation and requirements for site investigations for a tunneling project is the International Tunneling Association (ITA) report titled "Strategy for Site Investigation for Tunneling Projects" (ITA, 2015).

4.5 Compilation of relevant information and base map

The compilation of a base plan map (and profile if possible) based on existing information is a very useful tool that serves for the planning of site investigations. The base map should be based on recent topographic data as well as orthographic photographs of the project area and correctly referenced to any available survey benchmarks in the project area. A particularly useful tool is LiDAR topographic survey data that provides high resolution topographic definition that can often reveal the presence of geological contacts and faults, as well as landslides, particularly in mountainous terrain. Figure 4.2 presents an example of LiDAR to identify geological faults and fracture zones in bedrock.

Using the base map and a handheld global positioning system (GPS) it is possible to site practical and safe locations for proposed geological mapping programs, geophysical surveys, and drillholes.

The base map should include an initial presentation of the regional rock types in the project area along with major geological faults as indicated from national mapping information. Google Earth and/or aerial and photographs and satellite imagery should be also be used to evaluate the presence of any possible lineaments within the project area. A base map can also be draped over the topography using Google Earth to provide a useful tool for the identification of geological contacts and faults in relation to topographic changes. This three-dimensional tool is also useful for the identification of locations for field mapping especially with the use of a helicopter and identifying safe locations for landing.

4.6 Identification of key geological risks and possible concerns

As part of the planning of a site investigation program an initial evaluation of the key geological and construction risks and possible concerns should be performed by the designated tunnel design consultant in conjunction with the geological/geotechnical consultant. This initial evaluation can also be perceived as an initial risk assessment for the project to identify key risks based on previous experience in the project area or from other similar projects.

Key geological risks for a new tunnel project and site for possible concerns during construction may include the presence of the following:

* Major geological faults;
* Acidic groundwater;
* Very strong rock units;

Figure 4.2 LiDAR example for identification of geological faults.

- Very weak rock units;
- Non-durable rock units;
- Swelling rock units;
- High in situ stress regime, and;
- Low in situ stress regime.

   The presence of rock units that are non-durable and with swelling potential represent adverse conditions and are very high risk items to be confirmed early during the design stage in order that an appropriate investigation and testing program can be carefully planned to provide the necessary information in a timely manner for design.

4.7 Planning of phased investigations

The scope of the site investigations to be performed should be based on prudent practice and consistent with the design requirements for the tunnel project. As such, the scope of the site investigation program (starting with Phase 1) should be finalized by the tunnel design consultant in conjunction with the geological/geotechnical consultant after thorough consideration of all the available relevant information at the project site including historical information, consideration of site investigation programs for similar projects, and an initial evaluation of the key geological risks and possible concerns. Clients should be prepared that the first stage of a site investigation program often referred to as Phase 1, may detect conditions that were not expected and/or may have been limited in scope due to time or budget constraints. Under these circumstances it is necessary to consider an additional Phase 2 of investigations to target the unexpected conditions discovered in Phase 1 and also gap areas that were not addressed in Phase 1. For tunnel projects located in complex geological environments, it is not uncommon that three phases of site investigations are necessary to be completed in order to thoroughly evaluation all of the perceived risks.

While there exist no stringent guidelines for the scope and amount of work to be completed for a site investigation program for tunnel projects due to the high variability of subsurface conditions between projects, there are some useful practices to recognize, all with the common goal to investigate and address the key geological risks. The site investigation program should aim to evaluate all key risk areas and features with target drilling of suspected low strength rock units and moderate to major geological faults.

The amount of work to be completed as part of a site investigation program is subject to the complexity and constraints of the site conditions, the size of the proposed tunnel project in terms of tunnel size and length and the respective design requirements, the public profile of the project, and the proposed contractual risk allocation to be established by the client.

During the early stages it is important to investigate the identified key risks in order to confirm technical feasibility of construction in terms of practical and industry available technologies as well as a practical and acceptable schedule and estimate of cost.

If an appropriate amount of site investigations cannot be accepted as a project expenditure during the early stages prior to a major business decision to proceed, then an appropriate contingency for both design and construction risks should be included in the early cost estimate.

It is strongly advised that any proposed site investigation program should be subject to review by an independent technical review consultant engaged by the client.

4.8 Field mapping and ground proofing of inferred geological faults

The first step of a site investigation for a rock tunnel should be a comprehensive effort of field mapping. The field mapping should ideally be performed with a minimum of two persons to facilitate the collection of information and for safety reasons. The field mapping team should include the engagement of a specialist structural geologist or petrologist who has local experience and able to identify most rock types and any possible forms of alteration.

Mapping stations should be documented at most bedrock outcrop locations within the tunnel corridor as part of the mapping procedures with GPS coordinates, photographs, and representative samples at each station. Geological information that should be collected at each mapping station should include rock type, alteration type (if any), degree of weathering, orientations and nature of each fracture set, together with an initial evaluation of the rock mass quality using one of the typical rock mass classification systems to allow for an evaluation of potential squeezing.

As part of the field mapping program the ground proofing should be performed of all inferred geological faults previously identified from the desk study or inferred from Google Earth.

All efforts should be made to provide good access for the field mapping including the use of a helicopter to gain access to difficult locations or at high elevation where relevant bedrock outcrops may be present.

The base map should be updated following completion of the field mapping program and highlight the existence of any geological faults, inferred or confirmed, within the tunnel corridor. Perello *et al.* (2003) presents a critical review of geo-structural mapping methodologies for deep tunnels in mountainous areas which highlights the importance of the development of a reference geological model for a tunnel project. Figure 4.3 presents a significant geological fault of highly fractured rock among a large outcrop.

4.9 Geophysical surveys

Seismic refraction surveys should be performed at the proposed tunnel portal locations if significant overburden is inferred to be present in order to confirm the depth to bedrock for portal design purposes. Seismic reflection surveys, which allow for deeper penetration of energy pulses, should be performed over any areas along the tunnel alignment or corridor of suspected deep infilled glacial valleys and channels to confirm if such valleys or channels extend to the proposed tunnel elevation.

In remote mountainous areas where glaciers may be present, the use of ice radar should be performed across glaciers that overlie the tunnel alignment to evaluate the thickness or depth of the glaciers and if they extend to tunnel elevation for the design of the vertical alignment of the tunnel.

For long tunnels located in remote and difficult access areas airborne electromagnetic surveys should be performed for the identification of different rock types and geological faults between the rock units (Okazaki, *et al.*, 2011). Airborne electromagnetic surveys can provide information on rock quality and rock type to a depth of 100 m. An airborne electromagnetic survey was completed along the corroder of the proposed 36 km North Bank Hydropower tunnel in New Zealand that was located among steep forest covered mountainous terrain and helped to identify major geological faults within the corridor. Electrical resistivity surveys can however provide information about the subsurface conditions at a greater depth extending up to several hundred of meters (Martin & Farrukh, 2003, Rønning *et al.*, 2014). Due the high rock cover along the 6.4 km Alborz services tunnel in Iran, surface drilling was not performed and the site investigations were limited to an electric resistivity survey along the entire tunnel alignment that provided a good indication of potential high risk zones along the tunnel alignment of poor quality rock conditions. Figure 4.4 illustrates the results of the electrical resistivity survey and comparison to the geological profile for the tunnel (Wenner & Wannemacher, 2009).

Ice radar is a further useful geophysical technique that allows for the determination of the depth of ice associated with glaciers overlying bedrock and hence for the design of the vertical alignment for a tunnel. Glacier filled valleys are typically of a U-shaped

Figure 4.3 Example of geological fault in large outcrop.

geometry however the depth of the bottom is highly uncertain. Geotechnical drilling through glaciers is hazardous work due to the presence of crevasses and drilling challenges and hence ice radar offers a safer method for the evaluation of the definition of bedrock and represents confident data that can be relied upon for tunnel design.

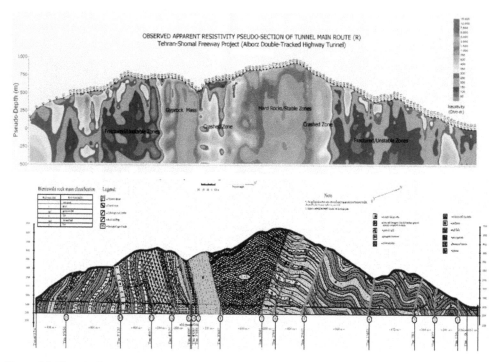

Figure 4.4 Electrical resistivity survey and comparative geological profile.

4.10 Borehole drilling

Borehole drilling is the most expensive form of site investigations especially for remote sites and therefore careful planning is required to minimize unnecessary incurred costs. The primary purpose of data collection for any site investigation program for a rock tunnel is to firstly obtain borehole core representing the geological conditions near the proposed elevation of the tunnel.

All boreholes should be planned in terms of priority of the required information for design purposes and recognizing that poor weather may impact the completion of the entire drilling program. The drilling program should be thoroughly planned and presented in the drilling program schedule presented as the number of drilling rigs, mobilization, set up and movement activities, and a practical and realistic drilling production rates depending if in situ testing is performed concurrently. The identification of practical and safe drilling locations can be evaluated with the use of high resolution orthographic photographs overlain into 3D topographic models as illustrated in Figure 4.5.

All rock drilling should be performed as rotary core drilling with a starting size of HQ (63.5 mm core diameter) using triple tubes (split tubes) in order to facilitate the recovery of all materials during drilling. If the drilling of a deep borehole experiences problems then it is acceptable to reduce the drilling size to NQ (47.6 mm core diameter) in order to attempt to complete the borehole to the target depth.

Figure 4.5 3D model of orthographic photograph overlain on topography.

All proposed boreholes should be drilled to a minimum of 10 m below tunnel elevation. All proposed boreholes should be located slightly off of the tunnel alignment to allow for instrumentation monitoring and prevent their intersection during construction. The drilling of deep drillholes should only be attempted using high capacity drilling rigs and include a sufficient set of spare parts particularly for work in remote sites. High capacity drilling rigs are capable of drilling depths up to 1000 m.

For boreholes that have been planned to target difficult geological conditions, such as fault zones, for the definition of their geometry and orientation, the boreholes should be drilled inclined in order to have a better chance of intersecting the contact of the geological fault. Upon confirmation of a contact of a geological fault by drilling, and if drilling is problematic to penetrate through the fault or within it to obtain core samples, subsequent boreholes should only be drilled vertically through the fault to reduce the risk of instability of the hole and possible collapse and jamming of drill rods and maximize the chance of obtaining core samples of the fault.

In general, most drilling planned to be performed along a tunnel alignment should be based on Inclined drilling of all proposed boreholes in order to collect additional data on rock mass structure and the possible presence of multiple rock units.

Horizontal drillholes drilled from portal locations can provide very valuable information on the expected geotechnical conditions for the early stages of construction. Figure 4.6 presents a horizontal drilling rig that completed a 1000 m long drillhole for a 1.5 km road tunnel in Hong Kong.

The drilling contractor should be required to maintain all important spare parts at the drilling site including spare triple tubes, bits and drilling muds in order to minimize any delays. In some cases access to drilling locations may only be possible by helicopter which requires greater safety awareness for all parties which include the restriction of

Figure 4.6 Long horizontal drilling rig.

drilling to daylight time only for helicopter flying In case of an accident. Figure 4.7 presents a drilling rig at a high elevation remote location accessible only by helicopter for the drilling of a deep 600 m drillhole.

All geotechnical logging of the borehole core, including photographs and sampling, should be performed at the drill rig to prevent the risk of impacting the quality of the core with transport.

4.11 In situ testing

The secondary purpose of data collection for any site investigation program for a rock tunnel is to obtain geotechnical and hydrogeological information on the rock conditions along the overall tunnel alignment but also within the overall tunnel corridor. For the design of rock tunnels it is important to gain some understanding on the spatial variability of the geotechnical and hydrogeological within the tunnel corridor and not to only focus the collection of information at tunnel elevation. For example the strength and permeability of rock commonly increases with depth but special geological circumstances, such as alteration, may have caused the weakening and loosening of rock at depth.

In order to gain an understanding of the spatial variability of the geotechnical and hydrogeological information it is necessary to perform in situ testing as a profiling approach along the entire length of most boreholes if time and budget allows.

Figure 4.7 Drilling rig at high elevation remote location.

It is important that in situ testing is performed in a sequence whereby the results of subsequent testing will not be impacted from previous testing. A typical good practice for the sequence of in situ testing is as follows:

- Fracture orientation survey (Borehole Televiewer);
- Rock Mass Permeability;
- Pump Tests;
- In situ stress testing, and;
- Borehole Deformability.

Borehole Televiewer surveys are commonly used to confirm the orientations of the main fracture sets. Care should be applied not to use a Borehole Televiewer below geological fault zones in boreholes where entrapment of the equipment can occur and possibly prevent subsequent testing to be performed. As noted previously, the entire borehole should be surveyed to provide information on the nature and orientation of fracture sets and any variation with depth.

The testing of rock mass permeability can be performed either concurrently with drilling using a single packer against the bottom of the borehole or after completion of the entire borehole to allow inspection of all of the core and the selection of the preferred testing intervals. Rock mass permeability testing should be performed along multiple sections of the borehole where there are limited fractures to represent the background or global rock mass permeability, as well as at discrete fracture zones or faults. All rock mass permeability testing should comprise five-stage pressure testing.

For some project sites where elevated groundwater tables are known to be present it may be warranted to perform pumping tests in order to allow for the evaluation of the effectiveness of groundwater mitigation measures during construction such as pumping and ground freezing.

For proposed tunnels with significant sections of moderate (> 300 m) to great (> 1000 m) depth or for hydraulic pressure tunnels it is necessary to perform in situ stress testing to provide an indication for design of both the minimum and maximum in situ stresses. Hydraulic jacking testing should be performed in boreholes to provide an indication of the minimum in situ stresses particularly along low cover sections of the tunnel alignment where there may be a risk of leakage for a hydraulic tunnel during operations. Hydraulic fracturing should also be performed to provide an indication of the maximum in situ stresses along the deepest sections of the tunnel where there may be a risk of overstressing of rock during construction.

Hydraulic fracturing and jacking testing can be performed within geotechnical drillholes typically up to depths of 500 m or from within shorter drillholes within underground excavations. Figure 4.8 presents hydraulic jacking testing to determine the minimum jacking pressures for a hydropower tunnel. Overcoring and slot testing requires access within an underground excavation such as an exploration gallery or existing tunnel. The most common method of overcoring is with the use of the Commonwealth Scientific and Industrial Research Organisation (CSIRO) hollow inclusion cell that allows for three-dimensional (3D) stress determination. Hydraulic fracturing and jacking testing is commonly a more rapid and less expensive testing method that can provide a larger set of testing data by profiling along a drillhole for an

Figure 4.8 Hydraulic jacking testing.

Figure 4.9 Overcoring in situ stress testing.

evaluation of in situ stresses in relation to depth and topography. The mini-frac hydraulic fracturing system of CSIRO available from www.esands.com is a low cost tool for two-dimensional (2D) stress determination. Overcoring however offers the measurement of the entire stress tensor and therefore is considered to represent a more reliable method of testing. Figure 4.9 illustrates overcoring testing within the power-house access tunnel during the early stages of construction of the Alto Maipo hydro-power project in Chile.

Finally, rock deformability testing can be performed in boreholes as a means to provide an indication of rock mass deformability of the various rock types along the tunnel alignment where there may be a risk of large deformation or squeezing during tunnel construction.

4.12 Selection and preparation of samples

The importance of protocols and procedures to be used for sampling for various laboratory testing for the design of rock tunnels cannot be over-emphasized.

The important aspects of sampling include the following:

- Representative selection including duplicate samples;
- Labelling;
- Documentation of Sample Lists;
- Protection and Packaging;
- Chain of Custody, and
- Transport.

Representative samples are required to be selected for testing of all rock types including varying degrees of alteration and/or weathering and/or durability in order to perform testing on the possible full variability of the anticipated conditions. A limited number of duplicate samples should be selected of key rock types for check testing at a separate laboratory as part of the quality assurance process and to have available multiple samples in case of damage during transport and preparation. All samples should only be removed from the core boxes after photographs of the core boxes have been completed.

All samples should contain correct and full descriptive labels either directly on the sample or on the form of protection wrapping presenting date, sample number, borehole number, depth, and type of testing to be performed. The location of where samples have been removed from core boxes should be marked similarly and if possible, replaced with a wooden or Styrofoam block of similar length. A zoom photograph should be made of all samples collected.

A sample list should be prepared and documented digitally that can be provided to all parties involved. All samples should be thoroughly protected with wrapping including plastic saran wrap, tin foil, bubble wrap or using simple plastic sample bags. Sample labelling should be included on all wrapping or sample bags. All samples should be packaged for transport in a box containing Styrofoam chips or bubble wrap for isolated protection of each sample to prevent any damage. Moisture or durability sensitive samples such as mudrocks (claystones, shales etc.) may warrant the preparation of each sample firstly in plastic wrap and inserted into polyvinyl chloride (PVC) sample tubes filled with melted paraffin wax.

Groundwater samples, subject to the type of testing, are typically required to be collected by accredited persons and only use new and fully sealed sample containers. Special care and safety equipment and clothing should be used for the collection of highly acidic groundwater samples.

Documentation presenting the Chain of Custody is important to the hand over and transfer of responsibility of the samples. All samples should be treated with great care for all handling for transport from the drill rig to a vehicle or other mode of transport from site to the laboratory. Samples should be well secured in a stable location where they will not be subject to disturbance. Samples should not be placed in the back of a pick-up truck where they may be subject to disturbance especially during transport from a remote site along unpaved roads. Available or empty core boxes should not be used for the transport of samples.

Figure 4.10 Rock block sample for testing of drillcores.

For deep tunnels where deep drilling would be required it may be appropriate to undertake sampling of representative rock blocks from surface outcrops of the various rock units along the tunnel alignment. These block samples can be collected as part of the field mapping program and should be labelled with a GPS coordinate or waypoint. These block samples can be drilled to obtain core samples for rock strength testing. For deep tunnels where easily accessible and representative outcrops are present it may be appropriate to obtain samples from short drillholes using a portable drilling machine. Figure 4.10 presents a typical rock block sample collected from a project site where rock cores have been drilled for laboratory testing. A further approach for the collection of representative rock samples can be achieved using a portable drilling rig and drilling short holes into outcrops as shown in Figure 4.11.

4.13 Laboratory and quality assurance testing

The results from laboratory testing of representative samples forms part of the key information for the design of rock tunnels and therefore the quality of the data is of great importance.

Laboratory testing of samples is typically performed for rock tunnels for the following:

- Uniaxial Compressive Strength (UCS);
- Brazilian Tensile Strength (BTS);
- Elastic Modulus and Poisson Ratio;
- Petrographic thin section analyses of mineral constituents;
- X-ray diffraction analyses of suspect infilling materials, and;
- Abrasivity.

Figure 4.11 Portable drilling rig for short drillholes for samples.

Additional specialized testing of rock samples is commonly performed for long tunnels where Tunnel Boring Machines (TBMs) may be used include the following:

- Drilling Rate Index (DRI);
- Cutter Life Index (CLI), and;
- Bit Wear Index (BWI).

Further specialized testing is typically warranted if moisture or durability sensitive types of rock are believed to be present and have been sampled including the following:

- Slake Durability Index, and;
- Swelling.

Where infilled rock fractures of potentially low strength materials are present it is important to perform direct shear tests to provide an indication of the shear strength of the fractures.

Laboratory testing should only be performed at accredited laboratories following standard industry procedures and should include a quality assurance inspection during the sample preparation and testing by an experienced member of the design team familiar with laboratory testing procedures. Photographs should be provided of all samples both prior to testing and after the completion of testing. Quality assurance tests should be performed at a different laboratory on a limited number of samples, particularly for uniaxial compressive strength, to confirm that all results are representative.

4.14 Field instrumentation and monitoring before construction

The completion of boreholes as part of a site investigation program allows for additional information to be collected regarding the groundwater regime along the proposed tunnel alignment.

Borehole drilling may be carried out on a continuous basis without stoppages (24/7 working hours) until the borehole is completed or may be performed only during daylight hours. Regardless of the drilling approach it is important to perform regular measurements of the depth of the water level in the borehole. Typically it is useful to perform such measurements at shift changes, or at the start of each day shift, after the water level has stabilized from the previous day of drilling.

In order to understand the expected groundwater regime and possible groundwater pressures to be anticipated during tunnel construction it is common practice to install piezometers in boreholes after the completion of drilling and in situ testing. Piezometers may include open standpipes commonly comprising PVC plastic tubes to allow the measurement of the water level in the borehole from all sections of the borehole or sealed at a particular interval of interest, for example, where high groundwater pressures may be present. Sealed piezometers can be constructed using a hydraulic pressure transducer or gauge that is installed down the hole within the interval of interest and backfilled and grouted.

After the completion of a site investigation program with the installation of piezometers it is important for the client to recognize that all of the piezometers should continue to be monitored during the design stage as well as during the construction stage and to appoint a designated party to be responsible for the monitoring and the reporting of the results. It is often appropriate to designate this responsibility to the tunnel design team unless the client has resources to perform this important ongoing data collection.

4.15 Pilot or test excavation/gallery

The construction of a pilot or test (exploration) gallery is commonly completed during a pre-construction stage or as part of a site investigation program for proposed large size tunnels, or just prior to advancing the main tunnel, or when a key risk has been identified during the early stages of the project. The adoption of this approach is particularly fully warranted when there is limited experience of the construction of such large tunnels in complex and/or the local geological conditions. Pilot or test galleries are typically constructed for traffic tunnels and hydropower caverns/chambers.

The construction of a pilot or test gallery provides the opportunity to learn about the excavation process and effectiveness, and excavation stability and behaviour, durability, and rock support design requirements for final design purposes. The location of the pilot of test gallery may be within the footprint of main works of the project or located in similar geology at a practical location that is approved prior to the main project.

The 5.5 km Piora Mulde exploration tunnel was constructed using an open-gripper TBM of 5 m diameter for the investigation of a sub-vertically oriented crushed formation prior to the construction of the 57 km Gotthard Base Rail Tunnel in Switzerland. A similar but longer 10 km exploratory tunnel was constructed using a 6.3 m diameter double shield TBM as part of the pre-construction investigations for the planned 55 km Brenner Rail Tunnel. Both these examples represent major exploratory tunnels but were warranted for the major rail tunnel projects in Europe.

Pilot or test galleries also provide the opportunity to perform additional testing of the ground conditions including in situ stress measurements by using overcoring techniques, deformation testing by using plate jacking, and general observations of excavation stability and durability using deformation instrumentation.

4.16 Reporting

The costs associated with the completion of site investigations warrants that a full documentation of all the relevant information including the performed field work, procedures, and the factual results are clearly and concisely reported for the use of the design team.

The site investigation reports commonly form part of the construction contract documents and are made available for the candidate tunnel constructors during the bidding process.

4.17 Drillcore photographs

Drillcore photographs represents an important record of the nature and quality of the geotechnical conditions. All drillcore should be photographed immediately during drilling at the site of drill rig before being transported to either a more suitable location for logging or storage. Drillcore is often sensitive to damage from transporting, particularly by vehicles along site access roads, but also by helicopter where the core boxes can be disturbed.

Drillcore can be photographed in groups of core boxes and should be performed in good quality natural light, or diffused artificial light, with no shadows. Photographs should be made of the core boxes oriented horizontally lying flat in a landscape format with the drilling depths starting from the top left and increasing to the bottom right. All drillcore should be photographed as dry core (non-wetted) and from standing vertically above the core boxes, typically requiring a short stepladder or stool, looking nearly vertical downwards. All core boxes should contain clear labels of the drillhole, date and box numbers, drilling direction, and all depths blocks within the core boxes should be clearly exposed and legible. Any drillcore losses should be clearly marked and preferably include a coloured spacer block of the length of the entire drillcore loss. A colour card and scale should also be included in the corner of the photograph. All drillcore should be photographed prior to the removal of laboratory testing samples and the completion of point load index testing. Figure 4.12 illustrates a good quality photograph of drillcore.

Drillcore that contains multiple zones of very weak materials including fault zone gouge should be photographed while within the split tubes in order to preserve the in situ appearance of the materials before possible damage during the placement into the core boxes.

Figure 4.12 Drillcore photograph.

4.18 Long term storage of drillcore

Drillcore produced from a site investigation should be well preserved and protected from damage and maintained at a designated and easily accessed storage facility for the duration of the design and construction of the project.

Multiple inspections of the drillcore can be expected to be required by the design team during the design and a mandatory inspection of representative drillcore should be part of the site visit during the bid stage of the project.

In the event that disputes have occurred during construction it is suggested that all drillcore continue to be maintained in case follow up inspections and further testing may be performed as additional information to resolve a dispute.

Chapter 5

Rock characterization

5.1 Regional and site geology

Comprehensive characterization of the geological, geotechnical, and hydrogeological conditions along the entire tunnel alignment or corridor is a fundamental requirement for the design of a rock tunnel.

The characterization of the geological conditions starts with the anticipated geological conditions in the project region and site. A careful evaluation of the regional geology is required to understand an appreciation of the large scale geological regime at the project which helps to define the overall risk profile of the project site.

The base map prepared as part of the planning for the site investigations should be expanded upon and updated based on the results of the site investigation program to include all relevant geological features including rock units, geological contacts between rock units, inferred and confirmed regional and local faults and their types, lineaments, geological fold axes, glacier boundaries, and groundwater springs.

A comprehensive geological plan map should be prepared and present all relevant geological information at an appropriate scale. Key reference locations including the proposed tunnel alignment should be presented on the geological plan map.

Tunneling in Rock (Wahlstrom, 1973) represents one of the earliest comprehensive publications that highlights the importance of the understanding of the various facets of geology including the petrography of unaltered and altered rocks for the design and construction of tunnels based on lessons learned from various tunnels constructed in the Colorado Rockies.

5.2 Tunnel alignment geology

The findings of the site investigation program field mapping and drilling should be considered for the compilation of a detailed geological plan map and profile along the proposed tunnel alignment. Both the geological plan map and profile should be presented on a single presentation of a drawing. A reference survey grid with a station chainage system along the proposed tunnel alignment should be presented on the plan map and profile respectively.

The interpretation presented upon the geological plan map and profile should be as consistent as possible however it is recognized that uncertainty may exist for the interpreted geological profile if complicated geological conditions are present along the proposed tunnel alignment.

All confirmed geological faults should be presented to extend to their known areal extents and all inferred geological faults should be extrapolated as dashed lines across the tunnel alignment and assumed to intersect the tunnel alignment on the plan map. All confirmed and inferred geological faults should be included on the geological profile at the respective interpreted locations along the tunnel alignment. The respective lengths of each rock unit should be defined in relation to the station chainage along the tunnel alignment as part of the geological profile.

The geological plan map and profile should be updated at each occurrence of new information for the project if additional site investigation data is made available.

5.3 Faults and fracture zones

Due to the typical impact of geological fault and fracture zones on tunnel construction it is important that they are correctly identified, documented with detailed geological descriptions (clay gouge, crushed rock, altered rock etc.) and presented clearly and in the correct locations on the geological plan map and profile. All geological faults that are identified and inferred within the tunnel corridor should be extrapolated across the tunnel alignment.

All geological faults and fracture zones should be assigned with labels such as F1, F2, etc. along the tunnel alignment. A definitive list of all geological faults and fracture zones with the inferred intersection length to the proposed tunnel should be presented and included on the geological plan map and profile. Geological faults can be defined with the following descriptive terms for rock tunneling projects.

Table 5.1 Type and characterization of geological faults.

Geological Fault Type	Anticipated Intersection Length, m
Minor	< 1
Moderate	1–10
Major	> 10

Geological fault zones represent key risks for tunnel construction. It is therefore desired to avoid geological faults if possible, especially large scale faults, by modification of the tunnel alignment. The feasibility of successful construction of a tunnel should be carefully evaluated based on the total amount of identified and inferred geological faults to be intersected, and in particular, the number of major faults expected to be intersected.

5.4 Rock mass fractures

Rock mass fractures have a strong influence of excavation stability for rock tunnels. Adversely oriented rock mass fractures in conjunction with weak shear strength due to associated weak geological conditions is important to be identified and presented for excavation and support design purposes.

The first step in the characterization of rock mass fractures is to identify the number of sets or families of fractures with different main orientations by presenting the data in a stereographic format. It is important to analyze all rock mass fracture orientation data on an individual basis (per borehole or outcrop) before any attempt is made at grouping of the data which can result in a false interpretation of the data and mis-identification of adversely oriented data. Accordingly, each individual source of data should be presented separately and evaluated. All data sources from borehole tele-viewer surveys should also be evaluated in terms of depth to identify if any significant changes of orientation occur with depth.

While the confidence of borehole oriented data has advanced to a high level, it is important to make a detailed comparison between outcrop mapping data to borehole data to confirm fracture sets, orientations and variability. The identification of signifi-cantly variable data sets should be recognized and labelled accordingly as separate structural domains along the tunnel alignment for design purposes.

Each identified family of rock mass fractures should be thoroughly described using appropriate geological terms (as per the recognized rock mass classification systems) including the presence of smooth surfaces and slickensides, weak infilling, and weak wall rock.

5.5 Rock strength

Rock strength can have a strong influence on excavation stability and support require-ments for rock tunnels, particularly deep tunnels where high ground stresses may be present.

Rock strength is typically described by the uniaxial compressive strength (UCS) and Brazilian tensile strength (BTS) and is based on the results of laboratory testing and/or field hammer estimation.

The results of all rock strength testing data should be critically reviewed with inspection of post-failure photographs to evaluate the type and representativeness of failure for acceptance within the data set. Intrusive rock types such as granites com-monly include micro-fractures which should be recognized as an inherent aspect of the rock strength.

The characterization of rock strength should be based on grouping of the data per rock type defined based on petrographic descriptors and degree of alteration. Hydrothermal alteration is commonly associated with rocks surrounding major mining areas and can be responsible for a significant reduction in rock strength.

The rock strengths of sedimentary and metamorphic rock types such as sandstone/shale and gneiss/schist are strongly dependent on the direction of testing with respect to the bedding and foliation and therefore should be evaluated on this basis.

Rock strength data should be presented using histograms for each group of data and include for as well as charts of depth versus rock strength for samples from boreholes in order to identify representative strengths at tunnel elevation.

Figure 5.1 illustrates an example of a sample for UCS testing that includes a micro-fracture that results in a false or lower than expected testing result of uniaxial compressive strength.

Figure 5.1 UCS rock strength sample with micro-fracture.

Point load index strength testing (PLST) is useful supporting information to char
acterize rock strength. PLST testing results from paired samples should be correlated to
the corresponding UCS testing results in order to determine a correlation factor to
apply to the expected large volume of PLST data. PLST data should be plotted in
conjunction with UCS testing data to provide confidence of the expected range of rock
strengths for various rock types.

5.6 Rock mineralogy

Rock mineralogy strongly influences rock strength and abrasivity.

The results of petrographic thin section analyses should be adopted for the correct
technical description of each rock type. X-ray diffraction analyses should be performed
for suspect infilling and/or other minerals that cannot be identified from petrographic
thin section analyses.

The amounts of hard mineral constituents including quartz, feldspar, epidote and
hornblende as well as soft mineral constituents including talc, calcite, and gypsum
should be evaluated and clearly presented for each main rock type along the tunnel
alignment to identify high risk mineralogy may impact tunneling conditions during
construction.

The presence of significant amounts of gypsum and anhydrite as well as clay and
other swelling types of minerals including smectites and laumontite is a common
concern for the design and long term operational performance of tunnels, particularly
for unlined hydraulic tunnels.

All petrographic thin section analyses should be performed by an accredited laboratory and well experienced petrologist.

5.7 Rock alteration

Rock alteration strongly influences rock strength. Rock alteration is commonly present in mining regions where there exist ore bodies of mixed economic minerals.

Rock alteration can increase the strength of rock due to silicification with the introduction of silica and can decrease the strength of rock due to the replacement of hard minerals with soft/weak minerals including sericite, chlorite, and clay minerals.

Rock alteration should identified and documented for all samples as part of the petrographic thin section analyses.

If strong alteration is present among the various rock types for the tunnel project as identified from samples from field mapping and petrographic thin section analyses, then a comprehensive evaluation of the extent of alteration with depth and along the tunnel alignment should be performed. A detailed review and analysis of the rock strength testing results may be warranted to fully understand the influence of alteration on the rock strength testing results. Figure 5.2 presents an example of hydrothermally altered granitic bedrock that has been discolored to a light pink and contains microfractures.

5.8 Rock abrasivity

Rock abrasivity influences the performance life of drilling bits for mechanized drilling jumbos, cutting picks for roadheaders, and cutters for Tunnel Boring Machines

Figure 5.2 Rock alteration sample – hydrothermal alteration.

(TBMs). Rock abrasivity is influenced by the presence of hard minerals including quartz, epidote, and hornblende.

Rock abrasivity is commonly defined in terms of the Cerchar Abrasivity Index (CAI) that is a standard laboratory test (ASTM D7625-10) that is influenced by the type of mineral constituents. Plinninger *et al.* (2003) discusses testing conditions and rock properties influencing the CAI value.

The results of rock abrasivity testing should be characterized for each rock type along the tunnel alignment and evaluated in relation to petrographic thin section analyses and rock strength testing to identify if any correlations are present.

5.9 Rock durability and swelling potential

Rock durability or deterioration and swelling potential is an important aspect for tunnel design in relation to excavation stability, initial support requirements, and the acceptability of unlined or partially lined hydraulic tunnels for long term stability and operational performance.

The deterioration of specific types of rocks commonly occurs as a result of changes in moisture content and swelling. The deterioration of rock can be observed during site investigations where core samples become exposed to increased moisture from natural humidity of the environment as well as undergo stress release or relaxation. Rock durability and deterioration and swelling potential is commonly associated with geologically young volcanic rocks including andesites, basalts, tuffs, and breccias, but also for low strength sedimentary rocks including clayshales and mudstones, where dissolvable and swelling minerals are common.

The deterioration potential or non-durability of any type of rock to be encountered along a tunnel alignment should be thoroughly evaluated and quantified and presented in terms of the severity of deterioration and swelling potential. Observations of rock deterioration, if apparent, should be documented with time dated photographs during a site investigation. An early indication of deterioration potential can be performed by simple soaking or immersion of samples in water and documenting observations over an initial period of time. In some cases it is possible to observe the rapid deterioration of rock from multiple inspections of drillcore during the period of a site investigation program. Figure 5.3 presents an example of early deterioration of young volcanic rock

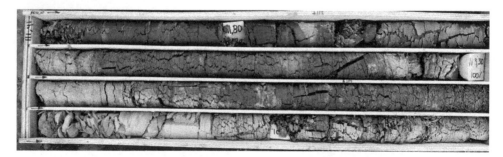

Figure 5.3 Early deterioration of rock core.

after 45 days during a site investigation program for a major hydropower project in Chile. Castro *et al.* (2003) presents the occurrence of expansive rocks with swelling minerals encountered during construction of hydropower tunnels in Chile and the design changes that were adopted.

Slake durability testing is a simplified test that should be performed to characterize the durability of suspect rocks. Swelling pressure testing should be performed on representative samples where swelling minerals such as clays, zeolites, and anhydrite have been confirmed to be present using the ISRM suggested method of testing (ISRM, 1999) in order that overly conservative testing results are not produced. Galera *et al.* (2014) presents an updated discussion on the swelling clay minerals of smectite, illite, zeolite and chlorite along with the risks associated with the design of hydropower tunnels from experience in Chile. Design recommendations of reinforced concrete linings including curved inverts are presented in relation to swelling pressures based on testing. Piaggio (2015) presents an appropriate testing methodology and characterization for non-durable and swelling rock units from the Andes based on the testing results from a comprehensive study including ethylene glycol and swelling tests. The correlation of properties and characterization of non-durable and swelling rocks can be complex and requires an extensive testing program to be performed.

The evaluation of rock durability should be performed in relation to the results of petrographic analyses and the identification of adverse minerals. Adverse minerals that are susceptible to deterioration include anhydrite, gypsum, zeolites (laumontite and leonhardite), and smectite clays. Basalts commonly include smectite clays due to deuteric alteration of primary minerals as well as amygdales filled with zeolites. The deterioration of highly amygdaloidal basalts by "crazing" or extensive micro-fracturing and expansion of these swelling minerals was experienced to a maximum depth of 500 mm in response to moisture and stress changes during the construction of the 45 km transfer tunnel as part of the Lesotho Highlands Project in the 1990s and warranted the complete concrete lining of the tunnel as a major design change (Broch, 2010). Figure 5.4 presents a petrographic thin section of a zeolite filled with laumontite, one of the most common swelling minerals and Figure 5.5 presents the deterioration of amygdaloidal basalt containing montmorillonite caused by natural humidity exposed during construction at the Lesotho Highlands transfer tunnels. Based on an extensive testing program both prior to construction and during tunnel excavation through the use of water spray rings, a critical threshold of about 30% by volume of deleterious minerals including 6% by volume of zeolites, can be expected to give rise to significant deterioration upon exposure to moisture which should be considered as part tunnel design (MacKellar & Reid, 1994).

The deterioration of andesites and other young volcanic rocks by softening can be observed during tunnel construction as a result of condensation from the tunnel water supply pipeline as occurred at the 21 km transfer tunnel at the Casecnan Multipurpose Project in the Philippines as shown in Figure 5.6 presents the deterioration of andesite due to condensation during construction of the transfer tunnel. Figure 5.7 presents another form of deterioration as dissolution of an anhydrite filled fracture due to repetitive direct contact from tunnel construction water from splashing.

Another risk of deterioration of non-durable rock is whereby the rock surrounding a hydraulic pressure is saturated for the first time when the original groundwater table is depressed. Under these conditions the swelling of young volcanic rock can occur and

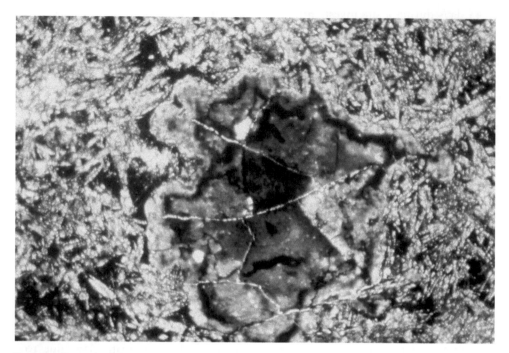

Figure 5.4 Petrographic thin section of zeolite filled with laumontite.

Figure 5.5 Deterioration of basalt during construction.

Figure 5.6 Deterioration of andesite during construction.

Figure 5.7 Dissolution of anhydrite filled fracture during construction.

Figure 5.8 Tunnel collapse of expansive volcanic rock after 9 years of saturation.

result in the increased loading of a shotcrete lining as occurred after 9 years of operation at the Rio Eoti Hydropower Project in Panama and resulted in multiple collapses along the 5 km tunnel alignment that required concrete lining of the entire tunnel to ensure safe long term operations. Figure 5.8 presents the large collapse of a section of the tunnel that was subjected to first time saturation.

5.10 Groundwater conditions, predicted inflows, and quality

The groundwater conditions prevailing along a tunnel alignment play an important role in both the design and construction of rock tunnels. Groundwater inflows can severely impact tunnel excavation and stability as well as worker safety. Groundwater inflows typically become mixed with construction water during tunnel excavation and support works and are required to be treated in accordance with local regulations prior to environmental release.

A proposed tunnel alignment may be located within a groundwater recharge or discharge area or cross the boundary of both recharge and discharge areas.

The groundwater conditions along a proposed tunnel alignment should be confirmed and evaluated in terms of static groundwater levels from long term monitoring of piezometers, effective groundwater pressures acting at tunnel elevation along the tunnel alignment, rock mass permeability in relation to depth, and environmental quality.

For deep tunnels where deep boreholes have not been completed it is appropriate to assume that very high groundwater pressures may be acting along the tunnel alignment equivalent to the entire cover above the tunnel.

The groundwater table will be drawn down during tunnel construction as the tunnel acts as a drain, and depending on the tunnel design approach, may result in a permanently reduced groundwater table for a "drained" tunnel design, or may re-establish itself to near pre-construction elevations for an "undrained" or commonly referred to as "tanked" tunnel design where the tunnel is fully sealed against long term groundwater inflows.

The volumes and pressures associated with groundwater inflows that are to be anticipated during excavation should be thoroughly evaluated and characterized. The accuracy of estimated groundwater inflows is not relevant but rather only the estimated magnitude presented as 10s liters/second, 100s liters/second, or 1000s liters/second.

Groundwater inflows typically emanate from geological faults and fracture zones and to a lesser degree from the surrounding rock mass around a tunnel. Realistic predictions of groundwater inflows can be performed using the method of Heuer (1995, 2005). Representative rock mass permeability values should be evaluated for all geological faults, fracture zones, and rock types from the results of in situ testing. Representative rock mass permeability values should be evaluated from in situ testing results in relation to depth within rock. Rock mass permeability may not necessarily decrease with depth in rock subject to in situ stress regime whereby a low stressed area can be associated with a relatively elevated rock mass permeability.

Predicted cumulative groundwater inflows should be presented for the envisaged approach of tunnel construction (for example with tunnel excavation from both portals or from a single portal) that will allow for an assessment of pumping requirements for all tunnel construction water. The source and storativity of groundwater within the rock mass is an important aspect to recognize for the potential for long term high volume and/or pressure inflows. The prediction of high volume and/or high pressure inflows at discrete geological fault zones or fracture zones allows for risk mitigation measures to be considered such as probe and drain holes.

The quality of the prevailing groundwater in terms of acidity and mineral constituents should be thoroughly evaluated to identify any risk of corrosion to rock support and influence to final tunnel linings. Environmental baseline reports should include relevant information regarding background groundwater quality.

Groundwater samples should be collected during construction and tested for acidity and mineral constituents. Any observations of corrosion to installed initial rock support should be documented and reviewed as part of the decisions for final support and lining.

5.11 In situ stresses

In situ stresses have a significant influence on excavation stability and rock support requirements that may involve both elevated and reduced or de-stressed levels of in situ stress.

Complex and elevated in situ stress regimes may be present in areas along global plate tectonics as well as in areas subject to glacial erosion where horizontal stresses are "locked-in". Most of the major mountain ranges around the world such as the Himalayas, Alps, Andes, and Rockies are associated with elevated in situ stress levels. Elevated levels of in situ stresses can also be expected to be associated with subtle topographic changes in terrain as well as along the toe of major mountain slopes oriented parallel to glacial valleys.

High or elevated in situ stresses, and in particular, respective in situ stress ratios present a risk of overstressing and increased instability for deep tunnels as well as moderate deep tunnels located in weak or low strength rock conditions. Similarly, a low stress regime or de-stressed area can adversely impact the stability of large size excavations and limits the contribution of the "clamping" effect that enhances the stability of fractured rock conditions, particularly for large excavations. Low in situ stresses may exist within topographic "bowls" or depressions as well as along major valleys due to relaxation or the presence of major geological faults. The presence of low in situ stresses can result in unacceptable leakage for hydraulic tunnels during operations through "open" or non-clamped rock mass fractures.

In situ stresses should be thoroughly evaluated and characterized as part of any site investigation for a proposed rock tunnel based on results from any of the well-established measurement methods including hydraulic fracturing and jacking, overcoring, or slot testing.

The World Stress Map Project (http://dc-app3-14.gfz-potsdam.de) provides an extensive database of earthquake focal mechanisms, well bore breakouts and drilling-induced fractures, in-situ stress measurements (overcoring, hydraulic fracturing, borehole slotter), and young geologic data (from fault-slip analysis and volcanic vent alignments) that may provide useful insight into the likely in situ stresses within a project area. All attempts should be made to measure the in situ stresses as part of a geotechnical investigation program either using hydraulic fracturing in drillholes or overcoring from within existing underground excavations in the project area.

The potential for overstressing should be evaluated based on consideration of in situ stress testing results in conjunction with representative rock strength data. Brox (2012, 2013a) developed an evaluation approach to characterize the potential for overstressing that can be applied for an entire tunnel alignment which requires an assumed in situ stress ratio as input for the evaluation. The risk of instability or leakage from hydraulic tunnels due to low stresses should be evaluated based on the results of hydraulic jacking testing and the respective minimum jacking pressures present between all identified sets of rock fractures. The testing results from both hydraulic fracturing and jacking testing should be compared to practical estimates of the theoretical overburden stress. For complex topographic geometry, three-dimensional (3D) stress analyses incorporating the topographic terrain should be performed.

Representative in situ stresses from site testing results should be considered as input for computational analyses of tunnel excavation stability and support.

The result of in situ stress measurements from all methods of testing should be plotted in relation to the depth of testing (ie. overburden stress) and/or in relation to the depth of testing from an existing excavation. Figure 5.9 illustrates an example of the results of hydro-jacking and hydro-fracturing testing. Figure 5.10 illustrates the result of overcoring testing in relation to testing depth and overburden stress.

While overcoring and hydro-fracturing testing provides the most reliable form of in situ stress results, hydro-jacking testing can also provide very useful indications particularly of an elevated in situ stress regime. The results presented in Figure 5.9 illustrate consistent results of minimum jacking pressures that are greater than the expected theoretical overburden stress and suggest in situ stress ratios approaching a value of 2.

5.12 Rock mass quality

Rock mass quality represents a useful characterization to present the variability of the geotechnical conditions of rock for tunneling purposes.

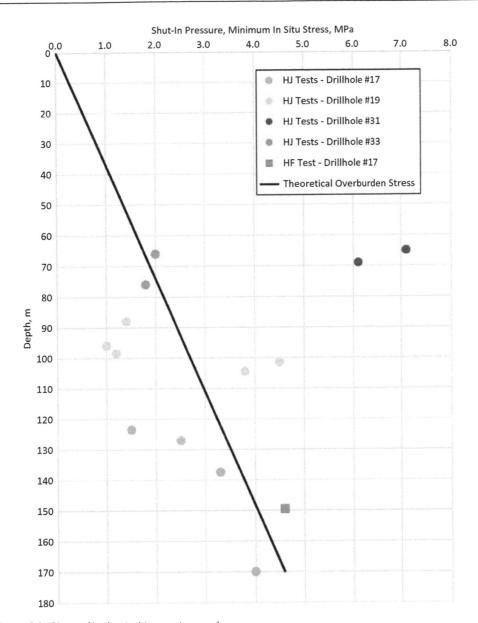

Figure 5.9 Chart of hydro-jacking testing results.

All borehole core available from the site investigation program should be logged for all pertinent geotechnical information using one or both of the internationally recognized rock mass classifications systems of the Norwegian Geotechnical Institute Q-System (Barton *et al.*, 1974) and the Rock Mass Rating (RMR) System (Bieniawski, 1976).

The Q and RMR rock mass classification systems are based on a series of parameters that describe the most important characteristics of rocks for their engineering properties. One of the earliest developed and easiest parameters to understand for describing

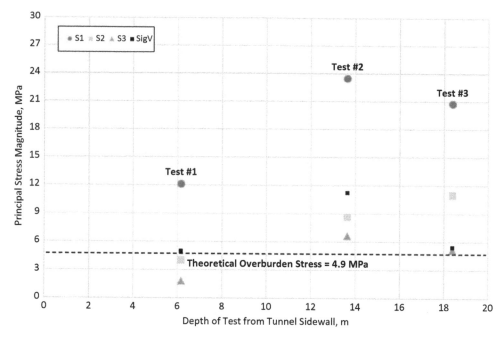

Figure 5.10 Chart of overcoring testing results.

the quality of rock is rock quality designation (RQD) that is presented as a percentage (Deere & Deere, 1988).

RQD is a measure of the degree of fracturing in a rock mass, measured as a percentage of the drill core in lengths of 10 cm or more. High-quality rock has an RQD of more than 75%, low quality of less than 50%. RQD is one of the key parameters included in the Q and RMR rock mass quality classifications.

All rock mass quality data should be evaluated and presented on an individual basis per rock type using histograms and per each borehole in relation to depth or elevation using scatter charts before any attempt is made at grouping of the data.

Ranges of rock mass quality should be presented for each type of rock and for each geological fault zone to be expected to be intersected at tunnel elevation along the tunnel alignment. Different methods of analysis can be performed to evaluate the distribution of rock mass quality including histograms for each drillhole, per rock type, per possible structural domains, per tunnel cover section, and in terms of the total grouped data.

During the early stages of design and site investigations when the tunnel layouts may not be finalized it is acceptable to perform rock mass classification on drillcore based on Q' and/or RMR' evaluations whereby the in situ stress groundwater components of the rock mass classifications systems are not incorporated in the overall assessment.

However, the classification of rock mass quality using only the Q' or RMR' parameters should not be used to represent geotechnical baseline conditions (GBR) as part of the conditions of contract since they do not reflect the entire in situ conditions for construction, and therefore cannot be simply compared to mapping conditions during construction. This practice can lead to significant claims during construction and should be avoided.

Both Q' and/or RMR' classifications must be completed for use in the GBR and this should performed once the preliminary design of the tunnel layout has been established and the groundwater and probable in situ stress conditions can be assumed from either in situ testing or best estimates.

5.13 Tunnel alignment and section characterization

Rock tunnels are linear infrastructure components and the characterization of the anticipated tunneling conditions should be presented based on the presentation of all the relevant geotechnical information including the following:

- Rock type;
- Geological Faults;
- Rock mass fractures;
- Cover/Overburden;
- Rock Strength;
- Mineralogy;
- Alteration;
- Abrasivity;
- Durability;
- Predicted groundwater inflows;
- In situ stress, and;
- Rock mass quality range.

The characterization of the tunneling conditions should be presented in conjunction with a topographic profile of the tunnel alignment that is referenced to a station chainage. Tunnel sections of similar characterization information should be identified and presented based on an evaluation of all of the information. These tunnel sections should be labelled as rock mass units (RMUs) or geotechnical domains (GDs) that provide a useful reference for design and during construction.

Chapter 6

Rock tunnel design

6.1 Design criteria and basis

The design criteria and basis or series of assumptions that are to be adopted for the design should be documented as per good industry practice at the start of the design stage. Some of the design criteria and basis of assumptions may be mandated by the client or national engineering and safety regulations and should therefore be strictly adopted. The design basis should clearly document the description of the tunnel component (e.g. tunnel size/ geometry, support), loading combinations, methodology, and the acceptance criteria (factors of safety). Additional information to be presented in the design basis should include technical standards and codes of practice, material properties, design parameters, and assumptions and considerations. It should be recognized that the costs associated with changes to design typically increase exponentially during the life of a project and therefore it is imperative to have clearly understood design criteria at the start of the project in order to advance the design to address all the required criteria.

The purpose of a proposed rock tunnel and the desired operational life of the tunnel should be confirmed with the client at the start of the design stage. Most tunnels in rock for civil infrastructure or hydropower projects should be designed for 100 years whereas mining tunnels typically have shorter lives but an increasing number of new mines are associated with extended lives reaching 40–50 years and therefore similar design philosophies should be adopted.

The design criteria and basis of assumptions should be regularly updated throughout the design stage as required with all new design information and changes to any of the original design assumptions and should remain as a "live" document throughout the design stage.

6.2 Technical standards and codes of practice

There does not exist any internationally recognized single series of technical standards or codes of practice for the design of tunnels in rock. Design approaches and philosophies have been developed around the world in many countries over the past few decades and in some cases differ from each other due to different levels of perceived safety by the governing authorities.

Some relevant examples of technical standards and codes of practice include the following:

- Swiss Norms SIA 197 and 198 – SIA (2004);
- Guidelines for Tunnel Lining Design – O'Rourke (1984);

- Geotechnical considerations in tunnel design and contract preparation – Hoek (1982);
- Design Guidelines for Pressure Tunnels and Shafts – Electric Power Research Institute – Brekke and Ripley (1987)
- Guidelines for the Design of Tunnels – International Tunneling Association (1988);
- Seismic Design of Tunnels – Wang (1993)
- Tunnels and Shafts in Rock – United States Army Corps of Engineers (1997)
- Guidelines for the Design of Shield Tunnel Lining – International Tunneling Association (2000);
- Road Tunnel Design Guidelines – Federal Highway Department Administration (2004);
- Austrian Guideline for Geomechanical Design of Tunnels – Necessity for Cooperation between Geologists, Geotechnical and Civil Engineers – Schwarz et al. (2004);
- Tunnel Lining Design Guide – British Tunneling Society (2004)
- Integration of geotechnical and structural design in tunneling – Hoek et al. (2008);
- The New Swiss Design Guidelines for Road Tunnels – Day (2008);
- Technical Manual for the Design and Construction of Road Tunnels – Civil Elements – Federal Highway Department Administration (2009);
- Best Practices for Roadway Tunnel Design, Construction, Maintenance, Inspection and Operations – National Cooperative Highway Research Program (2011), and;
- New Austroad Guidelines for Tunnel Design - Australian Tunneling Society (2011).

Additional tunnel design guidelines and practices may exist in other languages, and some of the above references may be updated with time. The reader is encouraged to search and review all applicable guidelines and practices during the early stages of tunnel design.

6.3 Tunnel cross section and internal geometrical requirements

The design of the tunnel cross section should be based on the internal geometrical requirements for safe and effective operations in conjunction with all geometrical requirements for practical construction, anticipated rock mass deformation, utilities, safety and operational components, and the final lining requirements for operations. Figure 6.1 presents an example of a tunnel cross section illustrating many of the required minimum clearance dimensions that define the overall tunnel profile envelope for design.

The internal geometrical requirements may be mandated from local government regulations and safety authorities and the current version of such requirements should be thoroughly reviewed.

The tunnel size required for unlined hydraulic tunnels should be based on a limiting maximum average flow velocity of 3.0 m/s for predominantly good quality rock conditions and lower limiting flow velocities for low strength or predominantly

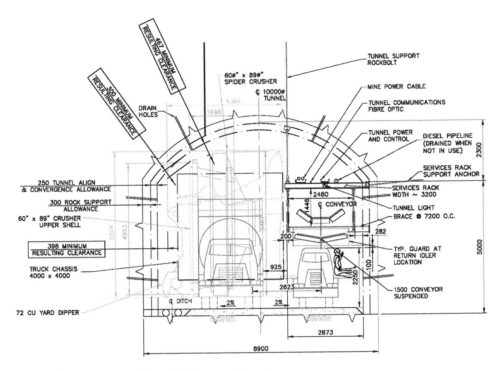

Figure 6.1 Tunnel cross section with clearance dimensions.

fractured rock conditions. Computational fluid dynamic (CFD) analyses should be performed to evaluate the distribution of flow velocity for a proposed tunnel cross section to confirm that the suggested threshold values are not exceeded.

6.4 Tunnel size and shape

The size and shape of proposed tunnels in rock should be based on the internal geometrical requirements as well as the construction space requirements. The size and shape of proposed tunnels in rock should importantly recognize to attempt to optimize and maximize tunnel stability for safe and practical construction including the installation of tunnel support as well as the use of practical size excavation equipment that does not significantly restrict appropriate production advances during construction.

A series of preliminary stability analyses should be performed, particularly for low strength rock conditions, to evaluate and optimize the size and shape of the tunnel in terms of excavation stability and the tunnel support requirements for practical construction, and to minimize risks during excavation. In some cases, for example in horizontally bedded sedimentary rock, the optimal shape of the tunnel may include a wide flat arch, which will assist to preserve the strength of the bedding in the roof arch area of the tunnel, thus allowing optimal tunnel support to be effective for stability. In contrast, for deep tunnels located in moderate strength rock subjected to high stresses, a sub-rounded tunnel shape can be expected to provide the optimal shape for stability and limit the extent of overstressing.

It may be appropriate to modify the size and shape of the tunnel cross section along the tunnel alignment subject to changes in the subsurface conditions. It may also be appropriate to increase the respective size of a tunnel based on a short notice availability of key tunnel construction equipment such as tunnel boring machines (TBMs) and drilling jumbos.

The size and shape of proposed tunnels in rock should also consider the possible influence and impact from adjacent structures while maintaining appropriate safe clearance separation from such structures.

6.5 Portal locations and support design

Tunnel portals are critical locations and the stability of these locations should be confirmed to be acceptable for the short term period of construction as well as the long term for operations.

Tunnel portals should ideally not be located in topographic depressions of gullies that may be subjected to flood events or within rockfall or avalanche runout paths. A comprehensive geohazard assessment should be completed for all proposed tunnel portal locations. If possible, tunnel portals should be located along or at the end of a topographic ridge or nose where good quality rock conditions are generally present. Figure 6.2 presents a tunnel portal sited among a small topographic ridge.

Figure 6.2 Tunnel portal in topographic ridge.

Tunnel portals should ideally be located within good quality rock conditions and where there exists minimal overburden to allow the construction of the portal to proceed without unnecessary delays and complications associated with the construction of complex excavation and support designs. Comprehensive stability evaluations should be performed for all portal locations to confirm the necessary support requirements.

In order to minimize the risk of flooding at a portal location it may be appropriate to elevate the portal location along a slope. Conversely, in order to facilitate tunnel grade requirements it may be appropriate to reduce the elevation of a portal and require construction of a recessed starting pit or small shaft.

The minimum rock cover at portals should generally be 1.5 times the tunnel width for fair to good quality rock conditions, and greater for poor quality rock conditions.

All portal locations should be thoroughly monitored to confirm their stability both during portal excavation as well during the early tunnel excavation period.

Portals for rock tunnels should be designed against rock mass failure including potentially unstable wedge blocks. Rock support for portals should include pattern rock bolts of sufficient length to stabilize all identified rock blocks that may become unstable due to planar, wedge, toppling, and rock mass forms of failure. Welded wire mesh in conjunction with shotcrete or chain link mesh should also be included to contain moderately to highly fractured rock conditions commonly present along the upper benches of a portal.

6.6 Horizontal alignment and separation

The horizontal alignment of a proposed tunnel in rock should be based on the requirement to align the tunnel over the shortest possible length. The horizontal alignment should also thoroughly consider to avoid any key construction risks such as the intersection of major geological faults, low strength rock units, and be oriented subparallel for extensive lengths to a main direction of geological features (faults, fractures zones, rock mass fractures) as well as rivers or streams. The minimum preferred radii is 25 m and 250 m for drill and blast and TBM excavated rock tunnels respectively. Sharp or acute turns or curves should also be avoided for hydraulic tunnels.

Multiple tunnels should be separated by an appropriate distance of at least three tunnel widths for fair to good quality rock conditions, and a greater separation distance for poor to fair quality rock conditions, especially for large size tunnels, to prevent any influence of the excavation of the adjacent tunnel. In addition to the lateral separation distance of parallel tunnels, the advancing face of each tunnel should be separated or staggered to also prevent any influence between the tunnels during excavation. The tunnel separation at the tunnel portals can be reduced to a minimum rock pillar if desired for operational requirements as long as special design measures are included to achieve and maintain adequate stability. Comprehensive stability analyses should be performed for multiple tunnels planned to be sited within poor to fair quality rock conditions where there is a risk of nearby influence. Several occurrences have been noted in the industry whereby the excavation of an adjacent tunnel has influenced the stability of a parallel tunnel.

Conceptual studies for tunnel projects should seek to identify all practical tunnel alignments that are constructible using available industry technology and meet the operational requirements for proper functioning. A comparative evaluation of tunnel alignments in terms of constructability with designated risk ratings and rankings should be performed as part of early studies in order to reduce the total number of tunnel alignments for the effective planning of geotechnical site investigation programs and to confirm all construction requirements for environmental approvals.

6.7 Vertical alignment

The vertical alignment of a proposed tunnel in rock should be based on the requirement to align the tunnel over the shortest possible length while respecting any limitations of gradient.

The vertical alignment should consider to minimize any key risks associated with the intersection of shallow weathered rock, and minimum rock cover, such as leakage for hydraulic tunnels, as well as maximum rock cover to limit the potential for overstressing.

The vertical clearance between a proposed tunnel and overlying existing infrastructure should be evaluated on a case by case basis considering the foundation design of the overlying existing infrastructure and the prevailing rock conditions. Detailed excavation stability analyses should be performed to evaluate predicted deformations associated with proposed clearance designs.

The vertical alignment of a hydraulic tunnel should be designed to prevent any negative transient pressures or only accept very small negative transient pressures acting along the tunnel crown. The design should be based on a hydraulic transient analysis in order to prevent water hammer and possible damage to the tunnel unless a surge facility is included in the design. The vertical alignment of hydraulic tunnels should be carefully evaluated in conjunction with hydraulic design considerations and analyses.

6.8 Practical grade

The vertical alignment of proposed rock tunnels is also subject to practical gradients that can be constructed in a practical and safe manner without unnecessary safety hazards.

The preferred maximum gradient for tunnels constructed using drill and blast methods is 12% to limit normal wear and tear on equipment as well as for safety considerations. The preferred maximum gradient for tunnels constructed using TBMs is 4%.

However, specialized equipment and procedures can be adopted to allow for the construction of rock tunnels at steep gradients including an inclined Alimak raiseclimber for steeply inclined drill and blast tunnels and starting cradles and secondary braking systems for TBMs.

The minimum practical gradient for hydraulic tunnels should be no less than 0.1% to facilitate flow conditions.

6.9 Intermediate or temporary access requirements

The construction of intermediate or temporary access adits is commonly adopted for long tunnels in order to allow multiple headings for tunnel excavation which results in the reduction of the total duration of tunnel construction. The location of all intermediate or temporary access adits should be sited in relatively good quality rock conditions to prevent delays of these adits since they typically represent critical path components for the overall project schedule. The locations of proposed intermediate access adits should also be based on attempting to optimize and reduce the construction schedule.

All or at least some of the intermediate or temporary access adits should be constructed to include bulkheads with practical sized access doors (typically minimum 3 m width) to allow for future access for inspection and maintenance.

6.10 Drainage requirements

The drainage design for rock tunnels is an important aspect that should be based on the hydrogeological conditions to be expected during tunnel operations. The amount of acceptable groundwater inflows into a tunnel is subject to the purpose of the tunnel where it is commonly desired to limit all inflows for all non-hydraulic tunnels (traffic, subway, conveyor etc.) in order to limit maintenance requirements during operations.

Uncontrolled groundwater inflows within a rock tunnel can result in unsafe operations due to ponding in recessed areas causing a slipping or skidding hazard as well as the formation of ice deposits during in cold climate.

All groundwater inflows, if not cut-off and fully prevented during excavation, should be fully channeled along designed drain ducts sized for an appropriate capacity of flow. Both radial drains along the tunnel profile and longitudinal drains along the tunnel invert are commonly incorporated into the tunnel design. For large groundwater inflows during normal operations it may be appropriate to include multiple sumps within the tunnel. An acceptable location for environmental release of all groundwater inflows should be included as part of the drainage design.

6.11 Invert requirements

The invert requirements for a rock tunnel are subject to the purpose and operating function of the tunnel. It is common practice to include a designed invert for traffic usage for most non-hydraulic and drill and blast excavated tunnels. A designed invert for drill and blast excavated hydraulic tunnels provides a significant benefit to operating hydraulics and reduces the risk of scour and erosion. Drill and blast excavated hydraulic tunnels where no designed tunnel invert is planned should at least be high pressure washed and cleaned with removal of all loose material and fines produced from excavation.

Designed tunnel inverts may include the backfilling and compaction of screened nature materials or crushed tunnel spoil or placed concrete. The volume of material required and effort required for placement for a tunnel invert is significant and should not be underestimated in terms of time and costs.

The construction of a final tunnel invert concurrently during tunnel excavation is a significant challenge particularly for drill and blast excavated tunnels due to the handling and control of construction water. However, the construction of a final tunnel invert incorporating the drainage requirements concurrently during TBM excavation can be practically achieved with minor impact to TBM excavation through the installation of pre-cast concrete segments or placement of concrete.

6.12 Operational design requirements

The operational design requirements for tunnels has steadily increased over recent years with an increase in the expected safety from governing authorities. The operational design requirements for tunnels is subject to the purpose of the tunnel, operating equipment within the tunnel as well as personal working or present within the tunnel during normal operations.

Tunnels that involve public use such as traffic, subway, pedestrian, bicycle, access, conveyor, and utilities are commonly subject to national safety standards which vary around the world. As with tunnel design standards many of these safety standard have originated from national road tunnel safety authorities and are frequently being updated based on lessons learned after accidents and overall improved safety practice and include the following:

- Standards for Road Tunnels, Bridges, and other limited Access Highways – NFPA 502 – 2014;
- Road Safety in Tunnels, World Road Association, 1996, and;
- Road Tunnel Safety Regulations – UK Ministry of Transport, 2007.

The majority of the operational design requirements address safety practice and requirements to be incorporated into any public use tunnel design and include ventilation, lighting, communications, emergency escape or refuges, safety bays, and fire suppression systems.

While it is recognized that the safety practice and requirements prepared by various national safety authorities are applicable for road tunnels, the various operational design requirements are considered to be applicable for special private access, conveyor and other utility type tunnels and therefore should be thoroughly reviewed and evaluated for any proposed tunnel design for human entry and public use.

A unique advantage of rock tunnels located in good quality rock conditions is that there exists a very low risk that any major fire incident may cause an instability to the tunnel and the potential for a full scale collapse and possible damage to any adjacent structures.

6.13 Access requirements

A consistent design aspect for the increased safety awareness to be incorporated into a tunnel design is the requirement for emergency and non-emergency access.

The emergency access requirements for public use tunnels is to allow any emergency services to reach an accident or incident location within the tunnel as soon as possible without delays and respond accordingly. Emergency access is commonly achieved with the addition of safety should lanes in traffic tunnels and dedicated emergency access and egress routes into subway stations.

As previously mentioned, intermediate or temporary access adits used for tunnel construction provide a practical solution to providing permanent access into tunnels for emergency incidents as well as inspections and maintenance.

Dedicated emergency access tunnels are being constructed as additional works for many existing public use tunnels around the world that did not include such access as part of the original construction. Such emergency access tunnels may include parallel small tunnels sized for emergency vehicles or dedicated access adits intersecting the main tunnel at key locations to reduce response times.

6.14 Design of hydraulic pressure tunnels

The design requirements for hydraulic pressure tunnels are very different for standard rock tunnels and are unique because the final loading imparted to the tunnel is only realized during operation when water under pressure acts against the internal surfaces of the tunnel. Figure 6.3 presents a historical perspective of the failure of unlined hydraulic pressure tunnels whereby the prominent mode of failure was fall out and either partial blockage or collapse due to poor rock conditions.

The cause of these blockages and collapses is that these specific locations within the tunnels were not adequately supported as part of the original construction. The common reason for the inadequate support is the non-identification of such adverse geological conditions during construction and the lack of understanding of the significance of such adverse conditions to tunnel stability in relation to the internal loading during normal hydropower pressure operations. Recent collapses of unlined hydropower pressure tunnels that occurred shortly after the start of operations at Glendoe in Scotland in 2009, Rio Esti in Panama in 2010, and La Higuera in Chile in 2011 typically resulted in

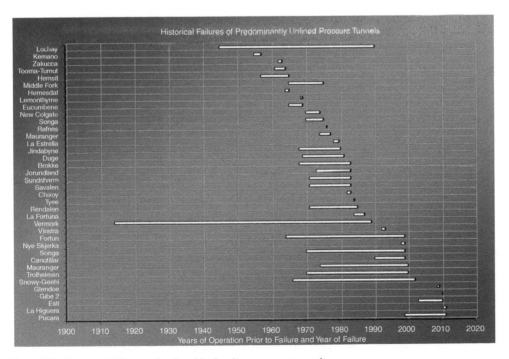

Figure 6.3 Historical failures of unlined hydraulic pressure tunnels.

loss of operations for more than 24 months and total costs related to lost revenue and repairs of $250 Million Dollars. These collapses highlighted the importance of thorough geological mapping during construction and the assessment of final design support requirements. The collapses at Glendoe and La Higuera occurred because of an under-supported geological fault that was identified and known about during construction. The reason for the collapse at Rio Esti is described later as a lesson learned case history.

Topography along the tunnel alignment and the hydrogeological conditions are the main design parameters to be carefully evaluated as part of the design. Benson (1989) presents the typical pressure tunnel design layouts as shown in Figure 6.4. The most common layout in terms of constructability, operating pressure risks to operations, and costs comprises a high elevation power tunnel of low gradient, inclined shaft, and underground powerhouse. Hydraulic pressure tunnels should avoid being sited under low cover topography over long distances, too close to the sides of major valleys, and parallel to major geological features where low in situ stresses may be present and result in significant leakage unless long steel liners are incorporated. Ground confinement analyses such as the Norwegian Cover Criteria (EPRI, 1987) should be performed to identify areas of limited confinement to allow appropriate modification of the tunnel layout to prevent leakage. Confinement analyses should be performed to evaluate both longitudinal and lateral or side slope geometry.

Where there exists extensive topography of low rock cover along the downstream section of the headrace tunnel alignment, and otherwise requiring an extended length of steel lining, it may be more appropriate to site the powerhouse underground at depth to maximize the total potential hydraulic head, introduce a tailrace tunnel, and thereby eliminating a significant length of steel lining. This type of design modification represents a trade-off that may introduce additional costs and may not be economically acceptable for a privately developed hydropower project.

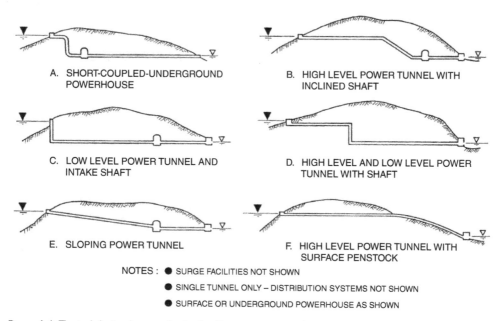

NOTES : ● SURGE FACILITIES NOT SHOWN

● SINGLE TUNNEL ONLY – DISTRIBUTION SYSTEMS NOT SHOWN

● SURFACE OR UNDERGROUND POWERHOUSE AS SHOWN

Figure 6.4 Typical design layouts for hydraulic pressure tunnels.

A basic design requirement for pressure tunnels is that the tunnel invert should be aligned at all locations over the entire tunnel alignment at least 5 m, or possibly more, below the hydraulic grade line corresponding to the maximum design flow to ensure that negative transient pressures, if developed during operations, do not influence or impact the long term integrity of the tunnel. A transient pressure analysis should be performed as part of design for a hydraulic pressure tunnel and the results should be thoroughly reviewed in conjunction with a hydraulic engineer. The impact of transient pressures can be relieved by incorporating a surge shaft but however this facility introduces a significant additional cost which may be prohibitive for some small hydropower projects.

The stability stages of a hydropower pressure tunnel can be defined as follows:

1. pre-excavation, pre-existing in situ stresses that are subject to local tectonics/geology;
2. Tunnel excavation with relaxation or overstressing and initial deformation;
3. Initial support and stabilization followed by possible further deformation subject to design and adequacy;
4. Final support either as additional support for unwatered stability due to unacceptable or ongoing deformation OR for scour and erosion protection – critical inspections required and review of detailed mapping to identify zones of weakness during hydraulic operations;
5. Watering up, re-establish groundwater regime, and;
6. Long term flow conditions subject to tunnel hydraulics and final lining with possible turbulent conditions and onset of scour with non-depressurization of pressures within fractures leading to erosion at discrete locations.

The typical failure modes associated with hydraulic pressure tunnels are as follows (Hendron et al. 1987):

• Excessive leakage due to high permeability rock or hydraulic jacking of fractures;
• Collapse and instability due to fall outs from rapid pressure fluctuations;
• Geological conditions susceptible to dissolution, deterioration, erosion, and swelling, and;
• Failure of linings due to buckling from external groundwater, poor contact grouting, and cracked concrete.

Hydraulic pressure tunnels are at risk of unacceptable leakage during operations if low in situ stresses are present to cause hydro-jacking of the rock mass around the tunnel. Low in situ stresses may exist below topographic depressions along a tunnel alignment, along the side slopes of major valleys due to post-glacial de-stressing, and near the downstream portal where there is limited rock cover. Steel liners are the standard industry design solution to prevent leakage from hydraulic tunnels during operations since steel is impermeable and the required length of a steel liner should be evaluated during the early stages of the design process due to the impact on construction costs and schedule. Merritt (1999) presents a simple graphic that illustrates the design logic and tunnel lining design requirements for different rock and hydrogeological conditions characterizing pressure tunnels as shown in Figure 6.5. While unreinforced and reinforced concrete linings are theoretical solutions for preventing excessive leakage, these design solutions are associated with the risk of construction quality and several failures have occurred in practice.

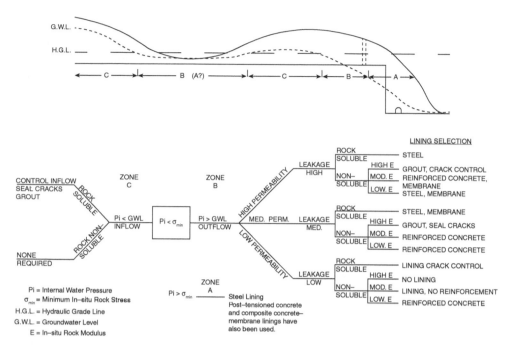

Figure 6.5 Design criteria for lining types for hydraulic pressure tunnels.

A preliminary assessment of the required length of a steel liner can be performed using the widely recognized Norwegian Criteria presented by EPRI (1987). Rancourt (2010) presents an updated approach for the preliminary evaluation of the length of a steel liner that incorporates the consideration of geological anomalies such as geological faults. Major geological features including faults, and shear and fracture zones represent possible de-stressing features that are critical leakage paths for hydraulic tunnels and should be clearly identified and characterized. The length of steel liners should extend beyond the locations of major geological features unless it is possible to confidently seal off these features by effective grouting.

Stress analyses should also be performed using computer software programs such as Phase2 (Rocscience, 2015) or FLAC (Itasca, 2015a) incorporating the topographic geometry along the relevant sections of the tunnel alignment. Figure 6.6 illustrates an example of a 2D stress analysis incorporating the variable topography along the tunnel alignment in order to provide an estimate of the length of steel liner required.

Hydraulic jacking testing should be performed at multiple locations near the end of the preliminary design length of the steel liner during the early stages of tunnel excavation. Multiple hydraulic jacking tests should be performed at each location to produce consistent results for the evaluation of the minimum jacking pressure within the prevailing rock conditions for comparison to the design pressure and final design location of the end of the steel liner. The thickness of steel liners is based on the internal operating pressure under static and dynamic conditions as well as the external groundwater pressure during tunnel dewatering. The design of concrete linings to prevent leakage and protect susceptible rock for both internal operating pressures and external

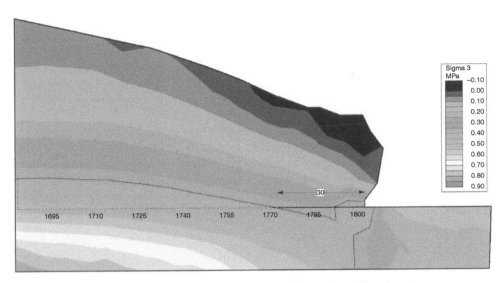

Figure 6.6 2D topographic stress model for preliminary estimate of steel liner length.

pressures is typically based on a load-sharing assumption utilizing the strength of the surrounding rock. The design process for such linings is complex and should be performed by a well experienced structural engineer familiar with the established design procedures in the industry.

Another critical design issue to be recognized at the onset of the design of a pressure tunnel is the intended method of operation of the hydropower plant by the owner which itself depends on the nature of the overall type and design of the hydropower scheme as run-of-river, peak loading with reservoir, and base loading. The intended type of power generation should be clearly defined and confirmed as part of the design criteria for a pressure tunnel.

If a power plant is planned to be operated as a peak power plant, that is to generate power to maximize commercial operations by producing during periods of high power prices, this commonly requires daily stoppages of the power plant, and resulting large pressure fluctuations within a pressure tunnel. This method of operation imposes much greater loading conditions on a pressure tunnel that must be recognized and evaluated and accounted for in the design of the tunnel support and final lining for safe and effective long term operations. The frequent stopping of hydropower plant operations results in ongoing pressure fluctuations imparted to the rock conditions around the profile of a pressure tunnel that can lead to the long term degradation of shotcrete support and linings. The deterioration of the tunnel can be further exacerbated by such ongoing pressure fluctuations if non-durable or swelling rock conditions exist.

The majority of hydropower pressure tunnels forms part of a run-of-river hydropower scheme and the operations of such tunnels are based on the available flow with very limited stoppages and do not impose additional loadings to the tunnel. In comparison, pumped storage schemes are operated as peak loading and are associated with frequent stoppages for the reversal of flow for pumping during low power demand periods, and therefore these tunnels are commonly designed with full concrete linings.

6.15 Seismic design considerations for rock tunnels

Tunnels constructed in rock are generally inherently stable under seismic loading conditions since no differential movement occurs along a tunnel with ground shaking. Accordingly, tunnels to be designed in rock do not have to consider any seismic effects. Portals will however be subjected to differential movement from ground shaking and therefore should consider seismic loading conditions in their design for the various forms of ground stabilization.

If a tunnel alignment can be expected to intersect a known, and active or potentially active geological fault, then rupture at the intersection should be considered as part of the design of the tunnel. The rupture of geological faults at the intersection of a tunnel can be accommodated within a design with the enlargement of the tunnel cross section to allow for differential movement at the intersection location. The size of any enlargement should also consider practical space requirements to perform emergency repairs and maintenance of the tunnel.

6.16 Constructability of design

Underground construction is associated with much higher risks in comparison to normal surface construction of infrastructure. It is therefore prudent to adopt a construction approach that seeks to minimize the risks of delays to the project schedule. The approach for the construction of a rock tunnel should be thoroughly evaluated as part of the tunnel design process as a series of constructability assessments.

Whenever possible, the overall approach for the construction of tunnels should consider excavation from both ends of a tunnel in order to reduce risks that may be encountered during construction thereby typically reducing the total duration and overall associated costs. However, in some cases it may not be possible to have access at both ends. Restriction or constraints may exist for some projects whereby the Client or governing authorities require that no work is performed from a particular end due to environmental concerns or community impacts.

The constructability assessments should confirm that the design is technically feasible for construction by utilizing currently available and proven technologies in the underground industry. The design of rock tunnels should not rely upon unprecedented construction approaches that introduces risks of cost and schedule overruns. The constructability assessments should also confirm that the desired or target construction schedule is achievable in order to attempt to meet the client's political or economic deadline. Alternatively, the construction schedule should be modified based on proven and achievable completion of activities and associated representative rates of production.

Where appreciable subsurface risks have been identified as part of the subsurface characterization and early risk assessments it is prudent to assume that multiple construction locations are necessary including work locations at both ends of a tunnel as well as possibly from intermediate access locations in order to address the risk of schedule delays. The construction approach for tunnels should not be based on aggressive or optimistic assumptions of rates of productions and/or geological conditions.

Chapter 7

Tunnel stability

7.1 General

The stability of a tunnel must be achieved and firmly maintained during each stage of excavation for worker safety, to prevent any partial or large scale collapse, and to allow for the installation of any required final lining.

The assessment of the stability of tunnels in rock has become increasingly simple with the development of user-friendly computer software programs to include as part of the design. The main challenge for these assessments is the evaluation and selection of representative rock parameters required for input into these software programs.

Tunnel stability assessments should be performed at each stage of tunnel design and regularly updated as new information becomes available particularly after the completion of a site investigation and for any changes to tunnel geometry. Tunnel stability assessments are required for a thorough understanding of the behaviour of the rock conditions during excavation and for the design of safe and practical rock support and linings.

7.2 Probable modes of instability

The stability of a proposed tunnel in rock may be affected by potentially unstable wedge rock blocks that can form around the periphery for prevailing fractured rock conditions, as well as by overstressing in the event that high in situ stresses are present at tunnel depth in conjunction with low to moderate rock strengths. Raveling and running ground may occur for highly fractured and non-durable rock conditions as well as in conjunction with groundwater pressure respectively. Squeezing conditions may occur at the intersection of very weak conditions such as those associated with major geological faults.

The probable modes of instability for rock tunnels includes the following:

- Wedge;
- Raveling;
- Running ground;
- Brittle overstressing/bursting, and;
- Squeezing.

A summary of the typical modes of failure in tunnels is presented in Figure 7.1 (Palmstrom, 1995).

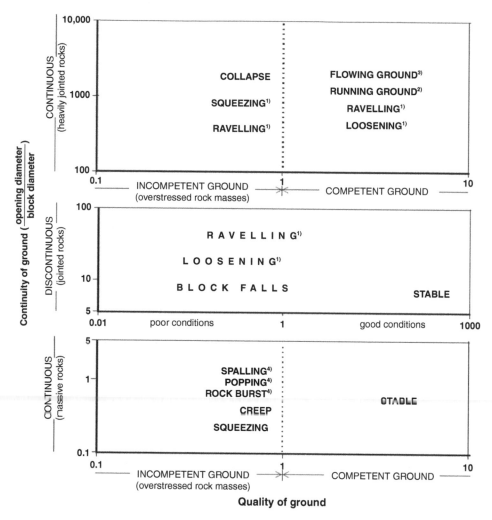

Figure 7.1   A summary of the typical modes of failure in tunnels.

7.3 Stability analyses and selection of parameters

Stability analyses should be performed for each type of probable mode of instability incorporating the proposed tunnel cross section geometry and representative geotechnical parameters for each separate geotechnical domain or rock mass unit.

The common forms of stability analyses for the probable modes of instability are presented in Table 7.1.

Representative geotechnical parameters for stability analyses should be selected based on a thorough review and evaluation of the characterization information as well as from consideration of information from similar projects located in similar geological conditions.

Table 7.1 Stability analyses for modes of instability.

Probable modes of instability	Stability Analysis
Wedge	Unwedge (Rocscience, 2015c)
Raveling	Phase2/FLAC (Rocscience/Itasca)
Running ground	Phase2/FLAC (Rocscience/Itasca)
Brittle overstressing/bursting	Overstress Analysis (Brox, 2013a)
Squeezing	Hoek and Marinos (2000)

7.4 Empirical assessments of stability

Empirical assessments of the stability of tunnels in rock provide a relatively rapid indication of the anticipated stand up time before support is required as well as a typical design in terms of capacity and extent of rock support.

The Rock Mass Rating (RMR) system from Bieniawski (1976) and the Norwegian Geotechnical Institute (NGI) Q-System from Barton *et al.* (1974) are the most common rock mass classifications that relate excavation stability to rock mass quality defined by the respective geotechnical parameters of each classification and also provides guidelines for rock support and an indication of stand up time before support installation.

Empirical assessments of stability based on historical case projects should only be considered during conceptual and preliminary design stages since the degree of conservatism included in the historical information is unknown and likely quite variable.

Notwithstanding the perceived limitations of using rock mass classification systems for an indication of excavation stability, this approach should be performed as part of the early stages of tunnel design to provide an indication of the likely range of rock support required and serves a useful purpose as part of the early design stage.

7.5 Kinematic stability assessment

The kinematic or structurally controlled stability of tunnels in rock is one of the easiest methods of stability assessment that should be performed. The UNWEDGE software program (Rocscience, 2015c) allows for a rapid assessment of the three dimensional (3D) stability of rock blocks or "wedges" that can be expected to form around the profile of a tunnel due to the intersection of the prevailing rock fractures.

Kinematic stability assessments require the identification and orientation of the main fracture sets which is most readily performed by presenting the fracture orientation data in a stereographic format such as with the Dips software program (Rocscience, 2015a).

Kinematic stability assessments allow for the recognition, in situ locations along the tunnel profile, mode of instability (sliding and gravity fall), and degree of stability of potential unstable wedges as well as the volume or size quantification of potential unstable wedges.

The UNWEDGE software program can be further applied to evaluate effective rock support designs for the risk of potential unstable wedges. Figure 7.2 illustrates a typical kinematic stability assessment. Kinematic stability assessments with the aid of available software should be regularly updated during tunnel excavation based on tunnel mapping data. Rock support designs should be regularly reviewed based on the results of these updated assessments and modified during construction if warranted.

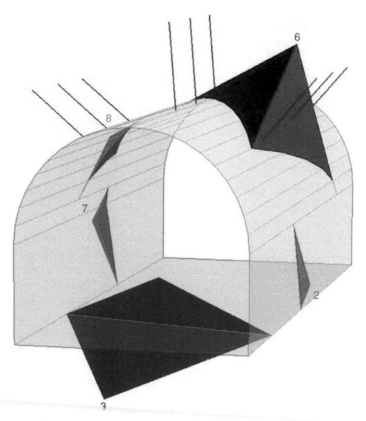

Figure 7.2 Kinematic stability of wedge blocks defined by UNWEDGE software.

7.6 Rock mass stability assessment

For weak, moderately to highly fractured rock conditions a rock mass stability assessment should be performed using one of the common industry software programs such as PHASE2 (Rocscience, 2015b) or FLAC (Itasca, 2015a) which incorporates representative geotechnical strength parameters and in situ stress conditions describing the rock conditions to be evaluated. These software programs incorporate the widely accepted rock mass failure criteria of Hoek *et al.* (2002) that can be described with input parameters based on geological observations using the Geological Strength Index (GSI) of Hoek *et al.* (1992). The results of such analyses are typically quite sensitive to the chosen parameters of rock mass strength and therefore a thorough evaluation of the rock conditions should be performed. The user-friendly nature of most software today also allows for sensitivity analyses to be performed with relative ease by considering a representative variation of parameters.

The presence of weak, moderately to highly fractured rock conditions commonly requires the incorporation of sequential or multi-drift excavation for medium to large size tunnels in order to maintain tunnel stability. The stability of medium to large size tunnels in very weak and highly fractured rock conditions typically requires the use of

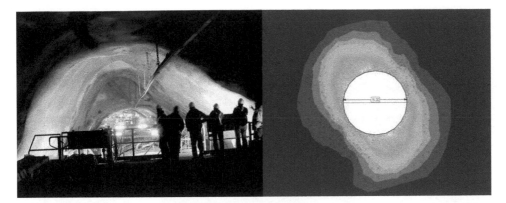

Figure 7.3 Stress analysis using Phases2D software.

small size sidewall drifts followed by further small drifts as part of the overall excavation sequence.

Figure 7.3 illustrates the application of the Phase2 software program for the back analysis and prediction of overstress and strain for the 14 m diameter Niagara hydropower tunnel sited at a relatively shallow depth of only 150 but in a high horizontal stress regime with low strength rock. Significant overstressing occurred during the construction of the tunnel and resulted in challenges for tunnel support and excavation. With the adoption of representative parameters the analysis was able to provide a good agreement with in situ observations of significant overstress.

Rock mass stability analyses should be performed during the early design stages and updated regularly with new site specific data in order to evaluate the sequential excavation requirements and the potential impact for tunnel support design and excavation. Rock mass stability assessments using common industry software programs and also importantly allow for the evaluation of any influence to adjacent structures such as overlying buildings or pre-existing underground infrastructure resulting from the proposed new tunnel construction. For complex excavation geometry and site conditions including sequential staggered excavation it may be necessary to perform three dimensional (3D) rock mass stability analyses to appropriately understand the 3D behaviour of the prevailing rock conditions.

7.7 Discrete element rock mass stability assessment

For widely to moderately fractured rock conditions it is appropriate to perform a discrete element rock mass stability assessment recognizing the actual geological conditions including the main fracture sets, discrete fracture zones, and geological faults in order to evaluate their possible influence on excavation stability.

A discrete element rock mass stability assessment should be performed using one of the common industry software programs such as PHASE2 (Rocscience, 2015b) or UDEC (Itasca, 2015b) which allows for the incorporation of actual rock mass fractures in relation to the proposed tunnel geometry. Figure 7.4 illustrates a two-dimensional (2D) discrete stability assessment using the UDEC software of multiple wide span road

Figure 7.4 Distinct element stability analysis using UDEC.

tunnels sited under shallow cover in weak and moderately fractured rock conditions and presents the extensive deformation as settlement along rock mass fractures that can be expected to occur prior to support. (Ghee *et al.*, 2011).

7.8 Evaluation of overstressing and characterization

An increasing number of tunnels are being constructed at moderate to great depths in rock and therefore can be at risk of overstressing which can have a significant impact during construction in terms of worker safety, excavation stability, and excavation advance. Overstressing is a behaviour of the failure of rock that is only applicable to brittle types of rock.

A simple and quick assessment of the potential for the occurrence of overstressing can be performed using the empirical spalling criteria of Diederichs *et al.* (2010) presented in Figure 7.5 that provides a relationship between the estimated depth of spalling and the ratio of the maximum boundary stress to the uniaxial compressive strength ($\sigma_{max/\sigma c}$). This approach suggests that overstressing as spalling can be expected to occur when $\sigma_{max}/CI > 1.0$ where CI is defined as the Crack Initiation Strength and typically equal to about 40% of the uniaxial compressive strength (UCS) and $\sigma_{max} = 3\sigma1 - \sigma3 = \sigma3(3k-1)$, where k is the in situ stress ratio.

The spalling criteria was compared to observations of varying degrees of overstressing for three deep TBM excavated tunnels in western Canada by Brox (2012) which allowed

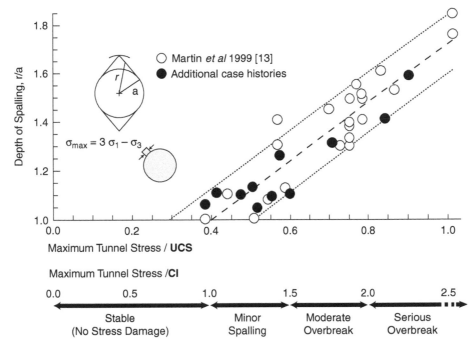

Figure 7.5 Empirical spalling criteria.

for the development of new approach for characterizing both the degree and the extent of overstressing along an entire tunnel alignment for varying rock and in situ stress conditions. Guidelines for overstress potential and rock support requirements were also presented.

Brox (2013a) further compared observations of varying degrees and extents of overstressing from several international projects as presented in Table 7.2 and provided validation of the new overstress characterization approach including for rockburst conditions which is important assessment for the evaluation of excavation methodologies. The overstress characterization assessment should be performed during the early stages of planning and design with limited rock strength data and assumptions regarding in situ stresses to provide a prediction of potential overstressing along the entire tunnel alignment. The assessment should be updated after the completion of site investigations with actual site data and form part of any risk assessment in order to evaluate the risks associated with various excavation methodologies.

While it is recognized that many tunneling practitioners would typically consider that long deep tunnels should be constructed using TBMs, an overstress evaluation that concludes a significant length of a tunnel alignment subjected to elevated levels of overstress and high risk, may rather warrant the use of high speed drill and blast excavation.

Figures 7.6, 7.7 and 7.8 present an example of the main graphic presentation to be produced as part of the overstress characterization assessment for the Seymour Capilano Twin Tunnels project where the method accurately predicted the onset of spalling and as well as the occurrence of rockbursting that occurred during TBM excavation (Brox, 2012). The first graphic presentation to create is the distribution of rock strength defined in terms of the uniaxial compressive strength (UCS) versus tunnel chainage along the

Table 7.2 Validation of predicted overstressing in deep tunnels.

No.	Project	Country	Year	Length, km	Size, m	Maximum Depth, m	Observed Overstress
1	Mont Blanc	France	1965	12	8.6	1400	Rockburst
2	Furka	Switzerland	1982	15	5	1400	Rockburst
3	Gran Sasso	Italy	1984	10	8	1500	Rockburst
4	Peheunche	Chile	1990	7	8	1400	Rockburst
5	Alfalfal	Chile	1990	8	5	1150	Rockburst
6	Lesotho Transfer	Lesotho	1990	45	5	1300	Severe
7	Rio Blanco	Chile	1991	11	6.5	1200	Severe
8	Kanetsu	Japan	1991	11	11	1175	Rockburst
9	Kemano T2	Canada	1992	8	5.7	650	Minor
10	Vereina	Switzerland	1996	21	6.5	1500	Moderate
11	Manapouri	New Zealand	2002	10	10	1200	Minor
12	Casecnan	Philippines	2002	21	6.5	1400	Moderate
13	Loetschberg	Switzerland	2005	34	8	2000	Rockburst
14	Parabati	India	2006	13	6.8	1300	Rockburst
15	El Platanal	Peru	2006	12	6	1200	Rockburst
16	Ashlu	Canada	2009	4.4	4	600	Moderate
17	Olmos	Peru	2010	14	5	2000	Rockburst
18	Jinping	China	2011	17	12	2500	Rockburst
19	Seymour Capilano	Canada	2011	14	3.8	550	Rockburst
20	Brenner Exploration	Italy	2012	10.5	6.3	1250	Rockburst
21	Cheves	Peru	2012	14	5	1400	Rockburst
22	Qinling	China	2013	28	12	2200	Extreme
23	Pahang Selangor	Malaysia	2013	46	5	1200	Rockburst
24	El Teniente	Chile	2014	9	10	1100	Extreme
25	McLymont	Canada	2015	2.8	5	800	Moderate
26	Neelum Jehlum	Pakistan	2015	13	8	1800	Rockburst

entire tunnel alignment. The second graphic presentation to create is the crack initiation strength, which is defined as 40% of the UCS, and the maximum boundary stress versus tunnel chainage along the entire tunnel alignment. The maximum boundary stress can be presented as a series of levels relating to different values of the in situ stress ratio, k, since this parameter is not commonly known during the early stages of a project. If the in situ stress ratio is known then it is appropriate to only use this single value. The third and final graphic presentation to create is the ratio of the maximum boundary stress to the UCS (σ_{max}/UCS) varying between values of 0.0 and 2.0, versus tunnel chainage along the entire tunnel alignment. Varying increasing levels of overstressing have been designated in terms of the expected amount of spalling as defined by varying ratios of the maximum boundary stress to the UCS.

Notable occurrences of rockbursts in deep tunnels include the Olmos Water Supply Tunnel in Peru (Lewis, 2009), the Brenner Exploration Tunnel in Italy (Grandori, 2011), and the Pahang Selangor Water Supply Tunnel in Malaysia (Kawata *et al.*, 2013).

The 13 km Olmos Water Supply Tunnel in Peru with rock cover of 2 km over a significant length of the alignment was constructed using an open gripper TBM. Tunnel

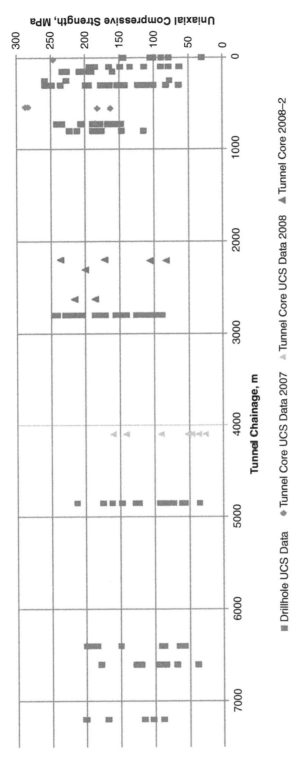

Figure 7.6 Rock strength (UCS) versus tunnel chainage.

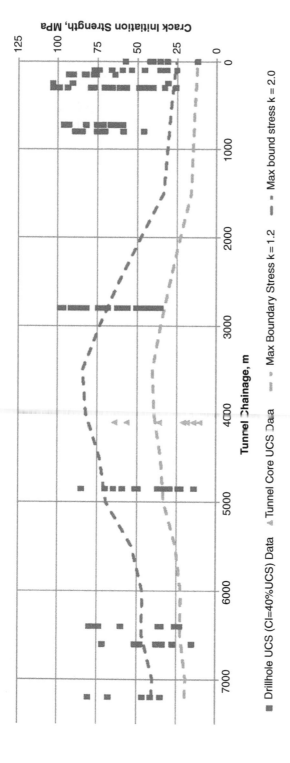

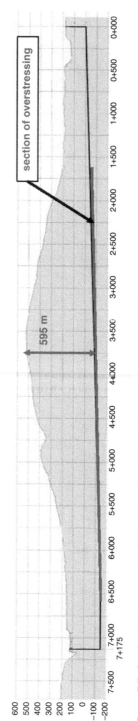

Figure 7.7 Crack initiation strength (CI) and maximum boundary stress versus tunnel chainage.

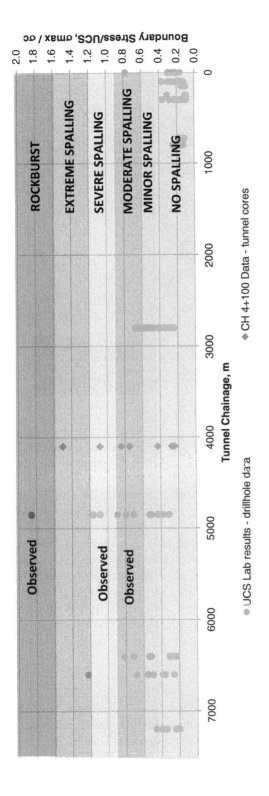

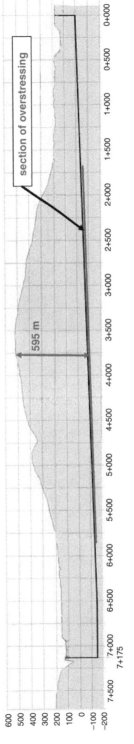

Figure 7.8 Overstress characterization in relation to tunnel alignment

Figure 7.9 Continuous overstressing along small diameter TBM tunnels.

construction experienced daily rockburst conditions under the high cover section and TBM excavation progressed slowly by incorporating de-stress blasting ahead of the TBM.

The overstress prediction approach correctly identified the onset of significant spalling and the impact to TBM gripper operations during construction of the Pahang Selangor water supply tunnel in Malaysia based on routine coring and strength testing of samples from the tunnel wall behind the advancing TBMs. Figure 7.9 presents an example of continuous overstress occurring under a rock cover varying from 400 m to 550 m that resulted in a significant increase of initial support requirements. Figure 7.10 presents an example of typical "dog-earing" shown by the breakout of rock along the roof area of the tunnel due to the presence of high horizontal stresses as confirmed from overcoring testing during excavation.

7.9 Tunnel stability at fault zones

Geological fault zones are present within most geologically deformed environments and represent high risk conditions for the stability of tunnels. Highly deformed major geological faults exist among many of the large mountain ranges including the Himalayas of Central Asia, the European Alps, the Andes of South America, and the Rockies of North America. The conditions associated with geological fault zones can vary significantly and include re-healed and silicified competent rock fragments, completely crushed rock into sugar consistency, completely weathered non-cohesive

Figure 7.10 Observed overstressing confirmed with in situ stress testing.

materials, and soft weak clay gouge. Geological fault zones are commonly associated with elevated levels of groundwater, and in some cases with very high pressures of 100s of meters. Fault zones can also act as groundwater compartments, which, upon their unexpected intersection during tunnel construction, typically results in a sudden movement of groundwater under high pressure and flowing materials. The unexpected intersection of geological faults zones can have a significant impact on the tunnel construction schedule with delays and cost overruns. Figure 7.11 illustrates the typical cohesionless materials associated with a major fault zone during the construction of the 7.2 km Vadlaheidi road tunnel in Iceland. Intersection of this fault zone resulted in a delay of more than 6 months and required significant grouting before excavation was resumed.

The intersection of geological fault zones during tunnel construction at acute or subparallel angles can greatly impact the stability of the tunnel over an extended length. Evaluation of the stability of geological fault zones at the intersection of a tunnel is very

Figure 7.11 Unstable materials at major geological fault zone.

challenging due to the typical highly uncertain conditions and geometry of the fault zones and the overall three dimensional (3D) geometry. High capacity tunnel support typically comprising lattice girders or steel sets in conjunction with shotcrete and possible pre-support in the form of spiling or forepoling should be considered to be required for the effective support of major geological fault zones for a distance of at least two times the tunnel width from the contact of the fault zone. The overall extent of the required high capacity support can be expected to be greater if the intersection angle with the tunnel is acute of sub-parallel. Deere (2007) presents some typical conditions associated with geological faults and describes some challenges for tunneling through geological faults. Some of the most challenging conditions associated with geological fault zones have been experienced with the construction of hydropower tunnels in the Himalayas (Carter *et al.*, 2005, Clark & Chorley, 2014). A very unique risk was experienced during the construction of the 26 km Gigel Gibe 2 hydropower tunnel in Ethiopia where a geological fault was intersected during TBM excavation and resulted in the reverse movement of the TBM under a pressure of 40 bars of mud flow (De Biase *et al.*, 2009). Another unique risk realized during tunnel construction was the occurrence of multiple rockbursts upon intersection of a major sub-horizontal geological fault at the Faido multi-function station of the Gotthard Base Rail Tunnel in Switzerland (Hagedorn *et al.*, 2008).

The stability and support design for the intersection of geological fault zones should be thoroughly evaluated as part of the tunnel support design based on the inferred conditions of each identified geological fault. The location, nature, geometry, thickness, orientation, and groundwater pressures of geological fault zones inferred to be present within a tunnel corridor should be thoroughly investigated and evaluated as part of a site investigation program.

7.10 Squeezing conditions

Squeezing rock conditions are generally described as large, time-dependent deformation and associated yielding and are a rare occurrence in tunneling. Collapses and fall-outs are not commonly associated with squeezing but may occur if significant deformation is allowed to occur without support. Squeezing conditions are commonly of limited extent along a given tunnel alignment and occur typically at discrete locations commonly associated with very weak rock conditions at geological fault zones under moderate to high rock cover where in situ stresses are significant. Common rock units where squeezing conditions may occur include shales, sandstones, schists, phyllites, clays, flysch, coal-seams and cataclastic rocks. However, in some cases, squeezing conditions can be present over much more appreciable lengths associated with a very weak rock unit under moderate to high rock cover. Squeezing conditions have resulted in significant delays and cost overruns for underground projects whereby re-profiling or re-excavation and support has been necessary to re-establish the design tunnel geometry. Squeezing conditions are relevant for the final design of a tunnel in terms of the stability of the tunnel profile but also the tunnel face. The deformation that occurs from a tunnel face under such conditions is typically much less but it may also become challenging for design and construction and require a multiple drift or sequential excavation approach with face stabilization measures. Figure 7.12 presents the severe squeezing and required re-profiling at the intersection of a fault zone at the multi-function station of Faido of the Gotthard Base Rail Tunnel under a depth of 800 m.

Figure 7.12 Squeezing conditions at TMZ Gotthard Base Rail Tunnel.

Squeezing conditions are commonly characterized in terms of strain defined as the maximum deformation in relation to the tunnel size in relation to the ratio of the estimated rock mass strength to the in situ stress (Hoek & Marinos, 2000). The prediction of squeezing conditions in terms of strain can be performed using the steps outlined by Hoek and Marinos (2000) by estimating the rock mass strength based on assumptions for the uniaxial compressive strength of intact rock, σ_i, the constant m_i, and the Geological Strength Index (GSI) and then applying the strain approximation formula defined as:

$$Tunnel\ Strain,\ \in\ =\ 0.2x\left(\frac{\sigma cm}{po}\right)^{-2},\%.$$

Where σ_{cm} is rock mass strength and p_0 is the in situ vertical stress. The rock mass strength can be determined by (Hoek & Marinos, 2000). Panthi (2006) proposes a simplified determination of rock mass strength for Himalayan rocks as:

$$Rock\ mass\ strength,\ \sigma cm\ =\ \left(\frac{\sigma i}{60}\right)^{1.5}$$

The formulations for the determination of squeezing conditions and the assessment of tunnel support requirements has been developed for use in the RocSupport software (Rocscience, 2016).

The results from an evaluation of squeezing can be compared to the squeezing characterization relationship (Hoek & Marinos, 2000) to assess the severity of squeezing as follows:

Table 7.3 Squeezing classes.

Squeezing Class	Tunnel Stability Status	Support Requirements	Tunnel Strain, ∈, %
A	Few Support Problems	Little to No Support	< 1%
B	Minor Squeezing	Simple Support	1% < ∈ < 2.5 %
C	Severe Squeezing	Heavy Support	2.5 % < ∈ < 5 %
D	Very Severe Squeezing	Yielding Support	5% < ∈ < 10 %
E	Extreme Squeezing	Yielding Support Essential	∈ > 10 %

This procedure can be simply applied to an identified rock unit along a tunnel alignment where squeezing is suspected or it can be applied along an entire tunnel alignment where multiple rock units may have the potential for squeezing. Kocbay et al. (2009) provides a useful graphic chart of the results of this approach for the characterization and quantification of the extent of potential for squeezing along the entire alignment of the Ermenek hydropower tunnel.

Some notable recent examples of appreciable squeezing conditions are presented in Table 7.3 (Barla et al., 2007, Hoek & Guevara, 2009, Kocbay, 2009, Mezgar et al., 2013, Agan, 2015).

Panthi and Nilsen (2007) confirmed this approach based on comparison to in situ convergence measurements for two sections of hydropower tunnels in Nepal. The possible consequences of squeezing in terms of damage to installed tunnel support

Table 7.4 Examples of squeezing conditions in tunnels.

Project	Length, km	Size, m	Depth, m	Geology	Strain, %
Lyon-Turin (SMLP)	2.4	9	425	schists	4.0
Kaligandaki	6	4.4	500	phyllites	8
Loetschberg	35	9	700	phyllites	5.5
Yacumba Quibor	24	4.5	1200	phyllites	18
Ermenek	8	6.6	475	Flysch+	10+
Faido MFS	8.6	9.0	1500	Gneissic fault	11
Sedrun TMZ	1.0	9	1000	phyllites	3
Uluabat	12	4	120	schists	6
Red Lake	6	4.2	1500	Talc-schists	7.5
Suruc	17	7.9	80	Marl/claystone	4.0

based on observations from Panthi and Nilsen (2007) and the author's experience of severe squeezing is presented in Table 7.4.

An example of the analysis of squeezing has been completed using the approach of Hoek and Marinos (2000) for the 14 km Vishnugad Pipalkoti Hydropower Tunnel as presented in Figure 7.13.

While squeezing conditions are commonly associated with geological fault zones at moderate to great depth, the presence of very weak rock units along a tunnel alignment under even shallow cover can also result in squeezing as occurred at multiple tunnels in Turkey. The evaluation of the behaviour of weak rock and its stability for tunnel design remains a challenge to understand if time dependent squeezing or overstressing can be expected. Martin et al. (2016) discusses and presents the behaviour and characterization of weak shale for three tunnel projects. Wirthlin et al. (2016) presents the squeezing conditions that were encountered during the construction of the 5.6 km New Irvington Tunnel and the ground supported that was implemented that included horseshoe-shaped steel sets along the entire tunnel as shown in Figure 7.14.

An evaluation of potential squeezing should be performed as part of early design in order to identify this potential hazard and any impact on design and construction. In the event that squeezing conditions are suspected during construction and were not identified as a possible risk during design, convergence measurements should be performed immediately and continue for each shift and be evaluated as soon as possible.

7.11 Stability of aging hydropower tunnels

Rosin (2005) presents a unique and representative quantitative risk based method for the evaluation of the stability of aging hydropower tunnels based on geotechnical conditions that provides an indication of the probability of failure or return period of a collapse.

This evaluation is based on accumulating the probabilities of failures due to identified instability or failure mechanisms that are deemed plausible based on consideration of the following key information:

- Tunnel history, geology, and support installed during original construction;
- Information from previous inspection reports and repairs;

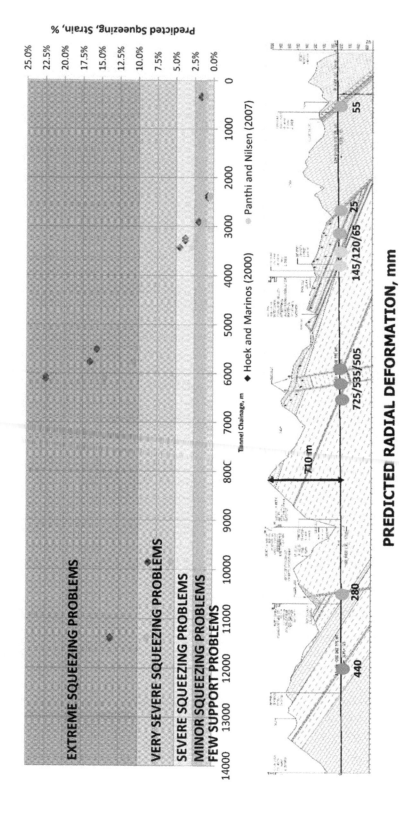

Figure 7.13 Squeezing characterization assessment.

Table 7.5 Consequences of squeezing to installed tunnel support.

Squeezing Class	Strain, %	Consequences
A	< 1	None, or very minor
B	1 – 2.5	Thin cracking of shotcrete, yielding of bolts
C	2.5 – 5	Initial buckling of steel ribs, breaking of faceplates off rock bolts
D	5 – 10	Wide cracking of shotcrete
E	>10	Severe buckling of steel ribs, significant damage to shotcrete and bolts

Figure 7.14 Steel ribs supports for continuous squeezing conditions.

- Identification of areas of concern based on tunnel mapping records, and;
- Review of case histories of other hydropower tunnel failures in similar geology.

The annualized probability of a tunnel collapse, P_A is defined as follows:

$$P_A = P_e \times (N_{RS} \times P_C)$$

where P_e is the annual probability of an initiating event, N_{RS} is the number of risk sites or locations where failure is deemed to be possible, and P_C is the conditional probability of failure at each of the risk sites. The selection of probabilities for uncertain events is based on the guidelines presented by Rosin (2005). This method of assessing the stability of aging hydropower tunnels is considered to be representative and can provide a plausible probability of failure based on realistic input from construction records.

7.12 Review of stability of existing tunnels in similar geology

The stability and performance of existing similar sized tunnels in similar geological conditions should be reviewed as part of any tunnel stability assessment. The historical performance of both similar sized tunnels and tunnels that were constructed in similar geological conditions can be invaluable information to consider for the correct under-standing of the behaviour of such rock conditions and for the appropriate design of the rock support and tunnel lining.

Site visits to similar tunnel projects of similar size and/or located in similar geological conditions should be strongly considered to be undertaken during the early design stage.

Chapter 8

Tunnel excavation

8.1 Practical considerations

The various methods of tunnel excavation that are commonly used in the industry including drill and blast, roadheader, and TBM should be thoroughly evaluated as part of any design for a proposed tunnel with consideration of historical information from previous projects constructed in similar geological conditions.

However, the method of excavation, and type of equipment, should be left to the selection by the preferred tunnel constructor and not be specified as part of the construction contract except under special circumstances where for example a particular method is perceived to be associated with high risks and is known not to have acceptably succeeded on previously similar projects in similar geological conditions.

The client and the tunnel design consultant are responsible to provide adequate geotechnical information for an evaluation of excavation methodologies. If high uncertainty remains to exist with regards to the risks associated with the geological conditions in order for all parties to agree with the most appropriate excavation methodology then additional site investigations should be performed such as the additional drilling of major fault zones or weak rock units.

Clients and tunnel design consultants should not be influenced by rumours in the industry of the poor performance and/or success of particular methods of excavation and should evaluate all relevant information and fully understand the reasons for the poor performance and/or success in order that a biased opinion is not formulated and the most appropriate excavation methodology can be implemented with consideration of all the project risks.

8.2 Minimum construction size

Although the construction industry has advanced significantly during the past few decades with the innovative technologies that allow for the excavation of large tunnels in rock there has not been a similar advancement of technologies for the construction of minimum sized tunnels in rock that are increasingly required for infrastructure.

The minimum construction size for tunnels in rock is subject to today's safe working practices including ventilation and practical equipment space requirements. In some jurisdictions the minimum acceptable dimensions are governed by safety authorities. Given these constraints and requirements the smallest size tunnels of long lengths is estimated to be limited to about 1.0 m in width and 1.5 m in height for the use of jack-leg drills with either hand mucking or slushers. While mechanized tunneling techniques such as micro-TBMs are applicable for the construction of small size tunnels less 1.0 m, these techniques continue to be limited to maximum tunnel lengths less than about 300 m and moderate strength rock conditions. The ongoing advancements in tunneling technologies can be expected to produce a solution for proposed small size and long tunnel in the not too distant future.

Many small hydropower projects are planned to require only minimum size hydraulic conveyances of less than 1.0 m in diameter with lengths in excess of 500 m. However, these projects are not yet technically feasible to construct at these limited sizes and lengths and therefore larger practical excavation sizes are required to be adopted whereby common equipment can be used for practical completion in a timely manner. The application of micro-tunnel boring machines (MTBMs) may be feasible for such projects subject to the nature of the rock conditions and access requirements. For strong rock conditions it is likely necessary to oversize even the minimum size for a MTBM application in order that cutters with adequate thrust can be used for practical and efficient production rates. For other cases where only drill and blast is to be considered, the typical minimum practical construction size is 2.5 m to 3.0 m which can be uneconomical for such cost-sensitive projects. In comparison, many civil infrastructure tunnel projects whose designs only require minimum size tunnels can accommodate the increased costs associated with oversizing.

8.3 Overbreak considerations

Overbreak in rock tunnel construction has been a common subject of dispute that has resulted in numerous claims and severe project delays due to significant reductions in excavation production due to larger mucking volumes for disposal and additional rock support.

Geological overbreak is commonly defined as the amount of rock unintentionally dislocated beyond the theoretical excavation line typically due to weak rock conditions and related geological instability beyond the control of the tunnel constructor. In comparison, over-excavation is commonly defined as the intentional excavation of rock beyond the theoretical excavation line by the tunnel constructor as approved by the designer.

Normal drilling and blasting practice in fair to good quality and moderately fractured rock conditions can typically result in over-excavation of 10–20% of the tunnel face area which is commonly accepted practice. However, poor drilling and blasting practice can result in over-excavation of more than 30% of the tunnel face area as shown in Figure 8.1.

Geological overbreak of more than 25% of the tunnel face area can be expected for poor quality or highly fractured rock conditions in conjunction with weak shear strength due to weak infilling and/or alteration along rock fractures. Geological overbreak, as well as over-excavation, can be reduce significantly and almost eliminated with controlled drilling and blasting practices including the use of electronic detonators as shown in Figure 8.2. The use of computerized drilling jumbos can commonly limit

Figure 8.1 Large overbreak along historical tunnel.

over-excavation to less than 10% of the tunnel face area in fair to good quality and moderately fractured rock conditions.

While the use of the nomenclature of "payline" shown beyond the theoretical excavation line on design drawings for contractual payment purposes has been adopted in the past for design and contract practice in the past, it is considered more appropriate to define the anticipated amount of geological overbreak as a baseline condition and include an estimated value in the Geotechnical Baseline Report prepared for bidders. Geological overbreak represents a construction risk that is a function solely of the geological conditions that should not be confused with over-excavation, and therefore should be included as a risk sharing item for contract purposes.

8.4 Drill and blast excavation

The drill and blast method of excavation is typically adopted for the construction of tunnels in rock of length less than 4 km for economic reasons. While drill and blast excavation for the construction of tunnels can certainly be used for good rock conditions, the method is generally preferred for variable rock conditions of poor to fair quality that includes numerous geological faults whereby greater flexibility exists to facilitate the installation of tunnel support.

Figure 8.2 Controlled drilling and blasting with electronic detonators.

Despite technological advances in drill and blast equipment including drilling penetration side dump buckets for scooptram load-haul-dump (LHD) vehicles, and bulk emulsion explosive loaders, the record production values for drill and blast methods have not increased significantly over the past few decades on account of increased safety regulations and practice. May tunnel constructors as well as clients and safety authorities now demand elevated safety practices from those used in the past that include a higher amount of minimum or safety support to be installed, limited concurrent activities at the working face, no concurrent drilling and loading of explosives, and flashcoat shotcreting of all exposed ground before scaling and installation of rock support. All of these practices have now resulted in limited and reduced rates of production compared to past performances.

Typical drill and blast excavation production rates for moderate size tunnels range from 4 m to 8 m per day, equivalent to one to two blast rounds per day, or one blast round per shift.

Drill round lengths of 4 m to 5 m are typically used for medium to large size tunnels with typically less than one tunnel width for tunnel sizes smaller than 4 m for fair and moderately fractured rock conditions. Drill round lengths should be limited to 2 m to 3 m in small tunnels to prevent the lack of confinement from causing excess damage to the tunnel profile and limit over-excavation. In many cases, drill round lengths greater than the tunnel width for small tunnels less than 4 m in width will result in very poor blast performance with choking of the blast and limited fragmentation due to a lack of free movement of the rock. For good quality rock conditions it is acceptable to use drill round lengths equal to the tunnel width up to the common practical maximum round length of typical equipment of 6 m.

The performance of drilling can typically be observed by the amount of half-barrels produced around the tunnel profile after the blast which should be documented as part of drilling and blasting operations for the ongoing optimization and improvement of practice.

Perimeter or smoothwall blasting practices should be adopted for poor to fair quality rock conditions with closely spaced blastholes, typically maximum 0.5 m, in conjunction with the use of decoupled, low density cartridge products to limit the shock wave and resulting damage to the tunnel profile.

The use of electronic detonators commonly results in lower energy absorption and hence less wall damage and result in smoothwall blasts with very limited over-excavation. Electronic detonators from a reputable supplier should only be used as quality control of these products remains a challenge in the industry.

Drill and blast excavation production rates are subject to tunnel size, rock conditions, type of equipment, support design, and experience of crews. For medium size tunnels with a width less than 6 m, production rates can be expected to range from 4 m to 8 m per day.

Typical powder factors range from 0.8 kg/m^3 to 2.2 kg/m^3 for tunnel sizes varying from 25 m^2 to more than 50 m^2.

For poor to fair quality and moderately to highly fractured rock conditions represent high risk conditions for scaling and all scaling of potentially loose rock should be performed as mechanical scaling using a hydraulic hammer to prevent the exposure of workers.

Drill round lengths should be limited subject to the quality of the rock conditions through specifications and defined by excavation classes with maximum allowable round lengths.

Probe drilling should be routinely performed in advance of all drill and blast excavation where there are risks of significant groundwater inflows and for situations where the geological conditions are anticipated to be highly variable.

Blasting vibrations should be limited when excavating in close proximity to adjacent or overlying structures or other concurrent works as well as young concrete and shotcrete less than 12 hours old.

A specialist blasting engineer should be engaged for all specialty drilling and blasting operations.

8.5 Blasting design

The use of standard drill and blast excavation results in vibrations emanated from the source of ignition of the blast. Excessive blasting vibrations can easily be transmitted to nearby existing buildings and presents uncertainty to the public about the safety of the ongoing works and in some case may result in damage to existing concrete and shotcrete.

Blasting should always be designed in terms of appropriate charge weights per blasting delay in order to prevent the exceedance of the stipulated maximum allowable peak particle velocity at the project site in order to attempt to prevent damage to existing adjacent and overlying infrastructure. If available, blasting monitoring data from previous work at the same project site or in a similar and local rock conditions should be reviewed to evaluate representative site specific blasting parameters to be used for the design. The presence of overburden overlying bedrock should not be assumed to dissipate blasting vibrations to acceptable levels.

Blasting designs can be developed to maximize excavation production for the site specific rock conditions by allowing long blast rounds to be effective in achieving full fragmentation and limiting instability to the excavation. Blasting designs include the use of an appropriate sized and length of a relief hole in conjunction with a burn hole geometry, production holes, perimeter holes, and charge weights.

Blasting rounds should be designed to limit the blast length in relation to the quality of the rock conditions in order to limit or prevent excess overbreak. Typically, the maximum length of a blast rounds is limited to the tunnel size/width for good quality rock conditions. However, the maximum length of blast rounds is commonly limited to 6 m with the use of standard drilling jumbo equipment.

While the practice of blasting design has historically been and continues to be the responsibility of the tunneling constructor, there is an increasing involvement of a blasting specialist, or well experienced blasting superintendent, through technical specifications on many projects, and particularly for rock excavation is sensitive areas adjacent to existing infrastructure.

8.6 Chemical rock breaking without vibrations

In some cases it may not be practically possible to perform controlled blasting of good quality rock in a timely manner without causing unavoidable excess vibrations when the works are sited in close proximity to existing sensitive infrastructure. For such special cases, the application of chemical expansion rock breaking products has been effective. Chemical expansion products are however limited in their ability to break rock in a timely manner and extended work periods should be recognized to be required to achieve good production. The use of chemical expansion products are typically used for small size excavation volumes or in very close proximity to highly sensitive structures such as large system computers for banks and financial institutions.

8.7 Scaling

Scaling should be performed after each blast in order to remove loose rock blocks that may have not dislodged as part of the blast. Scaling should be performed such that the scaling operator or crews are not directly exposed below the area of scaling. Tunneling practitioners should remain at a safe distance away during scaling.

Care should be taken when performing scaling in order to limit the amount of scaling which can result in the creation of loose rock blocks thereby causing increased instability.

Thorough scaling should be performed along the corners of the tunnel face and roof and sidewalls to remove any possibly loose rock blocks that are a risk during charging.

8.8 High speed drill and blast excavation for long tunnels

Several different technologies have been successfully utilized for high speed drill and blast excavation of long tunnels generally greater than 10 km in length. Several long hydropower, rail and traffic tunnels were constructed in Australia, Canada, France, and the United States between 1950 and 1980 utilizing rail-based access in conjunction with multi-level drilling platforms and high speed mucking machines. Some of these projects included the use of the Jacob's Sliding Floor (Petrofsky, 1987). The rate of advance of a single heading of these tunnels that ranged in size from 35 m^2 to 90 m^2 varied from 8 m/day to 23 m/day with an average of about 12 m/day.

A new form of high speed drill and blast technology was utilized starting in the mid-1990s at the southern half of the 22 km Vereina Rail Tunnel in Switzerland and was further used for portions of the 34 km Loetschberg and for the entire 15 km Ceneri Rail Tunnels also in Switzerland. The new technology developed by Rowa Tunneling Logistics (www.rowa-ag.ch) comprises the use of a large capacity mobile rock crusher and overhead hanging conveyor system positioned near the advancing tunnel face in conjunction with rubber tired drilling jumbo and side dumping scooptram. Sustained rates of production of 9 m/day to 12 m/day were achieved for a single heading for first time users of the overhead hanging conveyor system. The overhead hanging conveyor system that also houses the ventilation system typically requires a minimum sized tunnel of about 60 m^2 and is also considered to be most effective for tunnels of a minimum length of 4 km. Currently, the overhead hanging conveyor system is being used for the construction of a major road tunnel in India. The unique benefit offered by the overhead hanging conveyor system is the well organizing of all of the working activities and increased working safety due to free space availability along the tunnel floor.

The overhead hanging conveyor system is considered to represent the most technical and cost effective technology for high speed drill and blast excavation and is capable of application at remote project locations. Figure 8.3 illustrates the overhead hanging conveyor system for high speed drill and blast productivity.

8.9 Sequential Excavation Method (SEM) for weak rock

The excavation of tunnels sited in weak rock requires careful consideration and is most commonly based on the adoption of the sequential excavation method (SEM) which is also commonly referred to as the New Austrian Tunneling Method (NATM), (Thapa et al., 2013). Figure 8.4 illustrates the concept of the sequential excavation method for tunnel construction in weak rock (Urschitz & Gildner, 2004).

The SEM is typically based on multiple stages of small size drifts with restricted advance lengths before the next stage of tunnel support in order to maintain the overall stability of the tunnel. This method of excavation is also commonly associated with the excavation of a curved tunnel invert followed by immediate installation of tunnel support along the tunnel invert in order to provide full closure of the excavation profile. The excavation of the bench stage is the most critical and sensitive stage of excavation for large span tunnels in weak rock due to the significant re-distribution of stresses

Figure 8.3 Overhead hanging conveyor system for high speed drill and blast excavation.

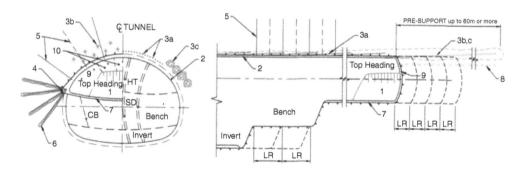

Figure 8.4 SEM/NATM concept layout.

within the haunches (Brox & Lee, 1995, Brox & Hagedorn, 1998). SEM commonly involves the installation of multiple types of tunnel support concurrently with excavation including standard rock bolts, shotcrete, and lattice girders but also as pre-support with forepoling/spiling as well as fibreglass rocks into the tunnel face as illustrated in Figure 8.5 (Hoek, 2000).

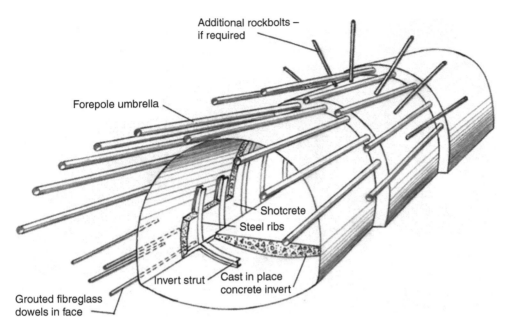

Figure 8.5 Tunnel support components for SEM/NATM tunnel construction.

SEM tunnel construction requires the implementation and reliance of geotechnical instrumentation and close monitoring in order to confirm the stability of all excavations at all times before allowing subsequent to proceed. This form of tunnel construction represents the highest risk for of tunnel construction which is exacerbated for large size tunnels sited at shallow depths in urban locations.

8.10 Tunnel Boring Machine (TBM) excavation

The use of TBM excavation for the construction of tunnels in rock has been successful for long tunnels with a variety of rock conditions for several decades. TBMs are commonly used for the construction of tunnels in rock of lengths greater than 4 km for economic reasons. Several tunnels in rock of shorter lengths have however been constructed with previously used TBMs. TBMs have also been used for a wide variety of special applications for the construction of tunnels in rock including inclined tunnels for mine access, hydropower pressure shafts, and metro station escalator tunnels. In general, TBMs should only be considered for the construction of tunnels when the majority of the inferred rock conditions are homogeneous, and of fair to good quality, and do not include more than 30% of poor quality conditions such as associated with geological faults.

The most common type of TBMs used for the construction of tunnels in rock are "open" or "gripper" and double-shield. Open type or gripper TBMs have been fabricated up to nearly 15 m in diameter and include a fingershield extending behind the cutterhead over the forward part of the TBM that provides limited protection of

Figure 8.6 Open gripper TBM.

workers but allows for the installation of rock support as soon as the rock becomes exposed before the position of the grippers. Figure 8.6 presents the largest size open type gripper TBM used to date of 14.5 made by the Robbins Company for the construction of the 10 km Niagara hydropower tunnel in Canada.

The effective thrust and overall advance of open type gripper TBMs may be significantly impacted by moderately to highly fractured rock conditions if unstable rock blocks become dislodged ahead of the gripper positions before they can be supported. A distinct disadvantage of the use of an open type gripper TBM is that tunnel workers are significantly exposed to the prevailing rock conditions. The use of an open type gripper TBM for a tunnel project characterized with a significant amount of overstressing (such as rockbursting) is therefore of very high risk and should be avoided unless special protection measures are included or special forms are implemented such as the McNally Roof Tunnel System™ for the workers. Open type gripper TBMs are most suited for homogeneous and good quality rock conditions and where a limited number of geological faults are expected.

Double shield TBMs include a three-component structural shield around the TBM that typically extends up to three times the diameter back from the cutterhead as well as grippers and rear thrusters. The multi-component shield allows for the advancement of the forward shield of the TBM while the rear shield remains fixed for gripping and is capable of achieving high rates of advance in challenging and varying rock conditions. Figure 8.7 presents the double shield TBM made by

Figure 8.7 Double shield TBM.

Terratec that was used for the construction of 12 km Xe Pian Namnoy hydro-power tunnel in Laos.

Single shield TBMs are used much less often for rock tunnels due to their actual lower advance rates in comparison to double shield TBMs. Single shield TBMs have typically been used in conjunction with pre-cast concrete segmental lining for the construction of water conveyance tunnels in non-durable rock conditions. The main advantage for single shield TBMs for such tunnels is the shorter construction schedule in comparison to the use of an open gripper TBM followed by a cast in place concrete lining. The distinct disadvantage of single shield TBMs for rock tunnels is that they are limited to thrusting off of the segmental lining only as they do not include grippers. As such, single shield TBMs are more difficult to become freed when entrapped within a geological fault zone. Single shield TBMs can operate in both open mode for relatively good quality rock conditions and closed pressurized mode in poor quality rock subject to high groundwater pressure. Figure 8.8 presents the single shield TBM by Herrenknecht that successfully constructed the 4.8 km tunnel under a maximum face pressure of 14 bars for the third intake at Lake Mead in the USA.

The distinct advantage of double-shield TBMs is the flexibility to install a variety of tunnel support including standard rock support comprising rock bolts, mesh and shotcrete as well as pre-cast concrete segments or steel ribs. The effective thrust and overall advance of a double-shield TBM may be significantly impacted by squeezing or high deformation rock conditions that may cause entrapment of the TBM shield. The use of a double-shield TBM for tunnel

Figure 8.8 Single shield TBM.

excavation through major geological fault zones should therefore be thoroughly evaluated with strong consideration of the implementation of mitigation measures prior to any such excavation. Under such difficult tunneling conditions double-shield TBMs can however utilize the rear thrusters to push off of pre-cast segments or specially designed steel ribs installed as part of the rock support that is typically designed for such conditions.

Open type gripper TBMs allow for the easy completion of probe drilling behind the cutterhead which may impact the advancement of the TBM. Double-shield TBMs allow for probe drilling through the rear shield with no impact to TBM advancement.

Open type gripper TBMs are ideally suited for proposed tunnels in rock that have been characterized with a significant portion of good quality rock conditions and limited amounts of geological faults. Open type gripper TBMs have commonly experienced delays due to highly fractured or blocky rock conditions that can cause clogging and blocking of the muck buckets that inhibits smooth operation. In comparison, double-shield TBMs are ideally suited for proposed tunnels in rock that have been characterized with a significant portion of fair quality rock conditions with substantial amounts of geological faults or potentially unstable rock conditions. Double-shield TBMs in conjunction with pre-cast concrete segments are also most appropriate for proposed tunnels in rock that are required to be fully concrete lined due to the risk of long term rock deterioration or aesthetic reasons.

In recognition of the challenges associated with the use of open type gripper and double-shield TBMs for the construction of tunnels with very difficult rock

Figure 8.9 Double Shield Universal (DSU) TBM.

conditions including geological faults and squeezing conditions, a hybrid type of TBM referred to as a Double-Shield Universal (DSU) TBM has been developed by SELI Overseas S.p.A in early 2000. The double-shield universal (DSU) TBM is a further evolution of the double shield TBM for the completion of the Val Viola water diversion tunnel in Italy (Concilia & Gandori, 2004). Further developments of the DSU TBM incorporate unique design features including very high torque, extreme overcutting, conical shields, steering controls, a flood door to prevent inrush of loose materials, and probe drilling ports for face treatment works. These unique features have allowed for the successful completion of several tunnel projects associated with very challenging and variable rock conditions that would otherwise have experienced significant delays or possibly not have been able to be completed by a standard double shield or single shield TBM (Grandori, 2016) Figure 8.9 presents the DSU TBM that successfully constructed 15 km of the Kishanganga hydropower tunnel in India under a maximum cover of 1400 m through medium strength sedimentary rocks.

TBM excavation production rates are subject to tunnel size, rock conditions, type of TBM and power and cutter size, support design, and experience of crews. The variation of advance rates for open gripper TBMs is presented in Figure 8.10 and highlights the large variability due to the different rock conditions and other parameters of the use of the TBMs for the various projects.

For medium size tunnels with a diameter less than 6 m, production rates can be expected for shielded TBMs in rock to range as presented in Table 8.1:

Table 8.1 Typical TBM production rates in rock – 6 m tunnel size.

TBM Type	Typical TBM Production Rate, m/day	
	Lower Bound	Upper Bound
Single Shield (with pre-cast)	8	18
Double Shield	12	20

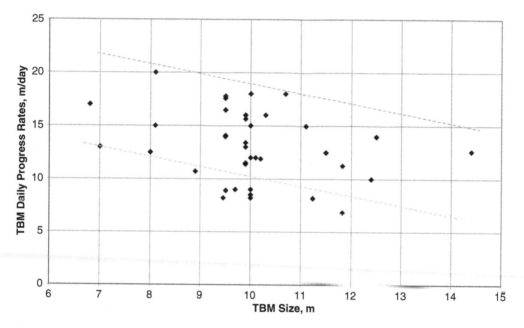

Figure 8.10 Variation of TBM open gripper advance rates.

The advance rates of TBMs should be thoroughly evaluated based on the first principles of penetration rates. The Norwegian Institute of Technology (NTNU) has developed a useful prediction model to provide estimates of open type gripper TBMs in various types of rock for different TBM sizes (Bruland, 1998). This prediction model has been developed into software and available at www.anleggsdata.no. The evaluation of TBM advance rates should include the review of historical project information of the use of TBMs of similar sizes in similar geological conditions. Simtunnel PRO 2.0 (Türtscher, 2016, www.simtunnel.com) is software that enables the prediction of TBM penetration and advance rates as well as cutter wear with the possibility to model all parameters as distributions and perform Monte Carlo analyses. Other methods for the assessment of TBM penetration rates include the Colorado School of Mines (CSM) Model (Yagiz et al., 2012), the application of Q_{TBM} (Barton, 1999), and the Rock Mass Excavability (RME) model (Bieniawski et al., 2007).

In order to determine the most appropriate type of TBM for the construction of a tunnel in rock a thorough evaluation should be performed of the entire geological conditions and possible construction risks during the early design process. This evaluation should be undertaken by tunneling practitioners with extensive experience in TBM tunnel construction and consider the construction of tunnel projects with similar geological conditions.

8.11 Assessment of TBM applicability

The applicability of TBM excavation for a given rock tunnel project is a commonly asked question from clients during the early planning and/or design stages of a project. The applicability of using a TBM for the construction of rock tunnel should be based on a comprehensive evaluation of multiple factors and project site information including tunnel length, site access availability, tunnel portal location and space, electrical power availability, labour experience, the nature of the geological/geotechnical and groundwater conditions and their distribution including rock strength, abrasivity and durability, and the number, nature, extent, and locations of geological faults along the tunnel alignment.

In general, the application of TBMs for a given project can be considered to be technically feasible for the following conditions:

- Long tunnels, with typically straight horizontal alignments with no tight curves, typical minimum length of more than 4 km, but can be less if good condition used TBM available;
- Low gradient of the vertical tunnel alignment, generally less than 4%, but steeper gradients can be constructed using special additional equipment;
- Adequate space requirements at the tunnel portal for the launch of the fully assembled TBM, but partial back-up launch can be accommodated with associated schedule delays;
- Adequate space requirements at the tunnel portal for mucking system either by mucking train tipping station or conveyor system;
- No availability of intermediate access adits due to terrain access challenges or environmental constraints;
- Majority of the rock conditions are characterized as fair to good quality with moderately fractured rock to facilitate cutting;
- Majority of the rock conditions are characterized as strong to very strong (75 MPa to 250 MPa) to limit overstressing and squeezing, and abrasive to very abrasive (CAI = 1.0 to 4.0) to limit excess cutter wear;
- Limited poor quality rock conditions (faults etc.), typically less than 20% of the total length of alignment and localized, not regularly or evenly distributed along alignment;
- Homogeneous rock conditions in relation to strength and abrasivity, limited variability for mixed face conditions to prevent irregular thrust and irregular cutter wear;
- Limited groundwater compartments of high pressure and volume to prevent flooding, inrush of flowing ground, need for extensive injection that impacts advance, and prevent good back filling if segments used;

- Electrical power supply from grid power or from nearby independent power station, and;
- Good TBM experience with local labour.

The use of TBMs for tunnel construction commonly requires the employment of well-experienced TBM operators, electricians, and mechanics. In some cases, the TBM manufacturer provides such specialist labour either for a minimum length of tunnel excavation during which training of local labour can be performed, or for the entire duration of tunnel construction. The costs associated with such specialist labour for TBM excavation are generally significant. In comparison, the costs associated with labour for tunnel construction by drill and blast methods are very modest and much less, particularly in developing countries.

In general, open gripper TBMs are acceptable for the excavation of competent, moderately strong and durable rock conditions that do not require extensive concrete lining but only partial shotcrete lining. In comparison, double shield TBMs are applicable for fair rock conditions including when there exists an appreciable amount of faults along the entire alignment. For hydraulic tunnels, double shield TBMs in conjunction with the installation of one-pass pre-cast concrete segmental linings are applicable for non-durable rock conditions where full or 100% concrete linings are necessary for long term operations.

While the applicability of TBM excavation for a particular project may be of interest to a client during the early stages of a project, it is important that, and common practice in the industry, that the method of tunnel excavation, including the type of TBM, is not specified by the client or the designer unless there exist special circumstances or critical constraints at the project site such as the prohibition of the use of explosives or the desire for optimum hydraulics for a hydropower tunnel. Common industry practice for tunnel projects is that the method of excavation is selected by the successful tunnel constructor.

8.12 The use of TBMs in squeezing ground conditions

The use of TBMs for the construction of tunnels in rock associated with numerous geological faults and/or low strength rock conditions characterized with squeezing is being increasingly contemplated simply due to the significant length of the tunnels under consideration and the assumed schedule benefits of TBMs.

The use of TBMs for the construction of tunnels in rock where squeezing conditions were encountered have typically experienced major delays due to entrapment of the TBM. The encountered squeezing conditions were typically associated major geological faults for the majority of these tunnels with some cases related to very low strength rock formations. In most cases, the occurrence and severity of the squeezing conditions were not identified and anticipated prior to construction as part of the characterization of the tunneling conditions. Table 8.2 presents a list of tunnel projects and actual delays experienced due to squeezing conditions. From the numerous case projects a typical project schedule delay due to squeezing conditions and entrapment of the TBM is about 180 days.

Ramoni (2010) provides an exhaustive discussion on the risks associated with the use of TBMs for tunnel construction through squeezing ground conditions. Terron (2014)

Table 8.2 Examples of squeezing conditions and entrapment of TBMs.

Project	Country	Year	Length, km	Size, m	Geology	Problems	Mitigation	Delay (days)
Evino Mornos	Greece	1995	30	4.1	flysch	squeezing	Pre-injection	150
Pinglin	Taiwan	1999	12.9	4.8	sandstone	Trapped (13)	Pre-injection	3000+
Yuncan	Peru	2000	9.0	4.1	granites	TBM trapped	bypass	180
Mohale	Lesotho	2000	16	4.9	basalt	infiltration	Pre-injection	120
Pont Ventoux	Italy	2000	13	4.0	schists	overstress	D&B	180
La Joya	Costa Rica	2006	7.9	6.2	lahar	squeezing	bypasses	285
Gigel Gibe II	Ethiopia	2006	26	7.0	volcanics	40 bar mud	bypass	180
Abdalajis	Spain	2007	7.1	10.0	volcanics	squeezing	Resin injection	120
Alborz	Iran	2008	6.3	5.2	andesite	squeezing	bypasses	540
Gerede	Turkey	2012	31.6	5.5	basalt	infiltration	Injection	500
Pando	Panama	2013	9	4.5	lahar	Groundwater	bypass	120
Kishanganga	India	2014	14.6	6.1	siltstone	collapses	Overcut/ inject	97
Kargi	Turkey	2014	118	10.0	volcanics	collapses	bypass	180
Tapovan	India	2014	8.6	6.6	schists	collapses	bypass	1000+

provides a comprehensive discussion (in Spanish) on the use of TBMs through geological faults with information from several case projects.

The use of TBMs for the construction of tunnels in rock where squeezing conditions may occur should be carefully evaluated to estimate the number, location and expected severity of the occurrence of such high risk conditions for acceptance within the project schedule or alternatively to confirm the requirement for mitigation measures to prevent such risks.

8.13 The use of TBMs for mining projects

There is a common misnomer in the mining industry that TBMs cannot be used however this is a completely false perception as the application is subject to geological risk.

There have been several successful and economic beneficial applications of TBMs for the construction of tunnels for mining projects. TBMs have been used for various purposes as part of new and expanding mining projects since the 1950s for new access, conveyance of ore and waste, drainage, exploration, and water diversion purposes. The use of TBMs for mining projects has not been without it fair share of challenges including mobilization of a TBM down a mine shaft and the attempted excavation of extremely strong rock. Also, many of the rock conditions associated with mining projects comprise highly altered bedrock that represents weak conditions that can pose special challenges for the use of TBMs. Brox (2013b) presents the technical considerations for the use of TBMs for mining projects.

Table 8.3 TBMs used for mining projects.

Project	Location	Purpose	Year	Length, km	Size, m
Steep Rock Iron	Canada	Access	1957	0.30	2.74
Nchanga	Zambia	Access	1970	3.2	3.65
Oak Grove	USA	Access	1977	0.20	7.4
Blyvoor	South Africa	Access	1977	0.30	1.84
Fosdalen	Norway	Access	1977	670	3.15
Blumenthal	Germany	Access	1979	10.6	6.5
Westfalen	Germany	Access	1979	12.7	6.1
Donkin Morien	Canada	Access	1984	3.6	7.6
Autlan	Mexico	Access	1985	1.8	3.6
Kiena	Canada	Access	1986	1.4	2.3
Stillwater EB	USA	Access	1988–91	6.4	4
Fraser (CUB)	Canada	Access	1989	1.5	2.1
Rio Blanco	Chile	Water supply	1992	11.0	5.7
San Manuel	USA	Access	1993	10.5	4.6
Cigar Lake	Canada	Access	1997	> 20	4.5
Port Hedland	Australia	Access	1998	1.3	5.0
Stillwater EB	USA	Access	1998–01	11.2	4.6
Mineral Creek	USA	Drainage	2001	4.0	6.0
Amplats	South Africa	Access	2001	0.35	2.4
Monte Giglio	Italy	Conveyor	2003	8.5	4.9
Tashan Coal	China	Access	2007	1.5	4.9
Ok Tedi	PNG	Drainage	2008	4.8	5.6
Los Bronces	Chile	Exploration	2009	8.0	4.2
Stillwater Blitz	USA	Access	2012–13	6.8(2)	5.5
Grosvenor Coal	Australia	Access	2013	1.0(2)	8.0
Oz Minerals	Australia	Access	2013	11.0(2)	5.8
Northparkes	Australia	Access	2013	2.0(2)	5.0

Table 8.3 presents mining projects where TBMs have been used for the construction of tunnels.

8.14 Minimum technical specifications for TBMs

The anticipated or planned use of TBMs for tunnel construction warrant the compilation of a series of minimum technical specifications by the tunnel design consultant in order to prevent the use of an inappropriate TBM for the tunnel project. Great care should be taken not to "over-specify" which would result in unnecessary requirements or exclude a particular TBM manufacturer. The typical minimum technical specifications for TBMs commonly address the following:

* Minimum and maximum TBM size;
* Minimum installed power;
* Maximum main bearing operating hours or new main bearing to be included with backup available within maximum time duration;

- Minimum cutter size or minimum thrust per cutter;
- Backloading cutters for medium sized TBMs;
- Minimum overcutting capability;
- Tapered shield geometry for shielded TBMs;
- Inclusion of fixed mounted rock bolting machines for medium sized TBMs;
- Inclusion of automatic steel rib erector arm;
- Inclusion of fixed mounted probe drill and probe drilling length for medium sized TBMs;
- Minimum length and radial extent of fingershield for gripper TBMs
- Minimum number and location of probe drill ports within forward and rear shields for shielded TBMs;
- Protection doors on mucking system;
- Enhanced thrust power for tailshield thrusters for shielded TBMs;
- Enhanced torque to facilitate freeing from entrapment for shielded TBMs, and
- Enhanced armouring of cutterhead face plates for highly abrasive rock conditions.

Additional specifications may include maximum procurement time, special factory testing requirements, and notification of shop testing visits by the client and consultant.

8.15 Roadheader excavation

Roadheader excavators have been used reliably for the excavation of moderate strength rock since their development for underground coal mining in the 1950s. Roadheaders are generally limited to the excavation of rock with a maximum uniaxial compressive strength of about 120 MPa and have been proven as best suited for massive, widely fractured, sedimentary rock that is generally of favorable stability to allow appreciable excavation stages before rock support.

Roadheaders are a flexible excavation technology that allow for the construction of irregularly shaped access tunnels, chambers and intersections. Excavation utilizing roadheaders requires the operation of enhanced ventilation system due to the greater amount of dust produced with the cutting process.

Roadheader excavation of small excavations can also be performed effectively utilizing roadheader cutting attachments connected to back excavators for the excavation.

Excavation with Roadheaders has been commonly associated with over-excavation of the required tunnel profile which has an important impact on the overall productivity, advance, and the construction schedule. Tunnel profile and alignment control software should be used to limit over-excavation whenever possible to maximize production.

Rock abrasivity is important to evaluate and estimate which impacts the wear of the drag picks and maintenance requirements and thus overall productivity. Roadheaders have been successfully used extensively in Australia for numerous large span road tunnels and intersections.

8.16 Methods for inclined excavation

Various methodologies are available for the excavation of inclined or declined tunnels oriented at gradients greater than typical for underground construction. Inclined tunnels are commonly required as part of hydropower projects and are referred to as pressure shafts. Inclined tunnels may also be required for utility and mine access or conveyor tunnels. Inclined tunnels can be excavated using drill and blast methods in conjunction with an Alimak raiseclimber which has been used at gradients ranging from 100% (45 degrees) to vertical. Inclined tunnels can also be excavated using open type gripper and shielded TBMs. Inclined access and pressure tunnels were recently completed using open gripper TBMs at the two major hydropower projects in Switzerland. Inclined access and pressure tunnels of 8 m diameter and 4 km in length, and twin inclined pressures tunnels of 5.2 m diameter and 1 km length were TBM excavated at the Linth-Limmern Hydropower Project at 24% (14 degrees) and 85% (40 degrees) respectively. These twin pressure tunnels excavated at a grade of 85% (40 degrees) represents the steepest inclination application for a TBM.

An inclined access tunnel of 9.5 m diameter and 5.6 km in length was TBM excavated at the Nant de Drance Hydropower Project at 11% (6 degrees). Inclined tunnels may also be constructed using shielded TBMs in conjunction with pre-cast concrete segmental linings as was completed for the 1.5 km long, 4.88 m diameter inclined pressure shafts at the Parabati hydropower project in India as shown in Figure 8.11. TBMs have been used for the excavation of steeply inclined pressures tunnels on a limited number of hydropower projects around the world.

Figure 8.11 Inclined pressure shafts with pre-cast concrete linings at Parabati.

Declined tunnels are commonly required for access into underground hydropower and mining caverns and more recently for access into underground metro stations. Drill and blast methods are not typically adopted for the excavation of steep decline tunnels due to the challenges of the removal of spoil. An example of the use of drill and blast methods for decline tunnels was at the Eastside Access Project in New York where multiple short escalator tunnels were constructed at 100% (45 degrees). Twin, 7.5 m diameter, 1.1 km decline tunnels were constructed using a hybrid shielded TBM at the Grosvenor Mine in Australia at 12.5 % (7 degrees).

8.17 Shaft excavation

The excavation of shafts in rock are typically carried out using drill and blast methods as top-down construction – referring to that the excavation advances in a downward direction starting from surface. This method is referred to as conventional shaft sinking. The size of shafts excavated in rock can vary widely subject to the intended purpose of the shaft. Many shafts constructed in rock are excavated in a circular shape in order to optimize the stability of the shaft and the rock support requirements.

Access shafts are commonly of limited dimensions ranging from 3.0 m in diameter to 6.0 m in diameter. Access shafts to facilitate the construction of a tunnel either by drill and blast or using a TBM are often excavated to larger sizes ranging from 8.0 m to 12.0 m in diameter to allow for the lowering of the equipment for tunneling. Figure 8.12 presents the 11.0 m diameter shaft that was excavated in rock to a total

Figure 8.12 Shaft excavation.

depth of 180 m to allow for the construction of twin drinking water tunnels using TBMs as part of the Seymour Capilano Water Filtration Project in Vancouver, Canada.

The equipment used for shaft excavation depends on the size and depth of the planned shaft. For most medium to large size shafts less than 200 m in depth for common civil infrastructure projects, excavation is usually completed using single or multi-boom crawler drilling rigs that move along the shaft floor along with mucking skips that are removed from the shaft with a crane located at the surface collar location. For most small size shafts greater than 200 m and commonly extending to 1000 m or more as required for deep mining operations, excavation is usually completed using a hanging gantry system that is suspended within the shaft as excavation proceeds downwards. The hanging gantry system comprises multiple levels for a hanging multiple-boom drilling jumbo, an arm loader or hanging scoop (Cryderman type) for mucking, and high speed mucking skips. Shafts constructed in rock are typically supported using standard rock support components and may be concrete lined subject to their intended purpose and design life.

Shafts in rock can also be constructed using the mechanized methods of raisebore drilling machines, Alimak raiseclimbers, as well as blindbore drilling machines. The use of raisebore drilling machines requires access at the bottom of the shaft and includes the drilling of a downward pilot hole followed by the connection of a reaming cutterhead that is subsequently pulled upwards to enlarge the shaft to the required size. Raisebored shafts are commonly constructed to diameters ranging from less than 1.0 m up to 8.0 m and for lengths ranging from 100 m to 500 m. The viability of shorter lengths is subject to project costs. Long raisebored shafts greater than 500 m require specialized survey alignment control equipment in order to reach the target location.

Alimak raiseclimbers comprise a hydraulically controlled platform that travels along a rack and pinion rail mounted along the rock surface of the shaft. Shafts are excavated using Alimak raiseclimbers by advancing upwards from the starting location at the bottom of the shaft and are commonly used for mining infrastructure including ventilation shafts, orepasses, and escape/egress ways. Alimak raiseclimbers are typically used to construct small size shafts ranging from 3.0 m to 5.0 m and for lengths ranging from 100 m to 500 m. Shafts are constructed using Alimak raiseclimbers typically at inclinations from 60 degrees to vertical. Standard drill and blast methods are employed for rock excavation and traditional rock support is installed for stabilization.

Blindbore drilling machines can be used for the construction of shafts in rock for low to medium strength rock conditions only. Blindbore drilling machines have typically been used to construct shafts ranging from 4.0 m to 6.0 m in diameter and for lengths ranging from 100 m to 300 m. Blindbore drilling machines advance downwards excavating the full shaft diameter and have been mainly used in competent rock conditions because support and/or lining of the shaft is only performed after completion of all excavation.

Shafts constructed in rock using mechanized methods are typically more stable and require less rock support because the surrounding rock is not subjected to damage from blasting. The use of mechanized methods for the construction of shafts is increasing for a variety of applications as part of underground infrastructure for civil engineering projects. In general, the stability of shafts can be expected to decrease with increasing depth as the excavation is subjected to higher in situ stresses in relation to the strength of the encountered rock. Shafts excavated in excess of 500 m can be expected to be subjected to appreciable in situ stresses than may cause failure of rock to a level up to rockbursting. The excavation of shafts at great depths therefore poses unique risks to workers and requires appropriately designed support systems to stabilize the exposed rock conditions and for the protection of workers both during excavation and future operations.

8.18 Cavern excavation

Caverns may be constructed for a variety of infrastructure requirements including underground parking, hydropower, underground bulk storage, water reservoirs, desalination plants, refuge transfer, oil and gas storage, sewage treatment, and security storage (explosives and ammunition). Rock caverns represent an innovative solution for storage particularly for key infrastructure desired to be located close to or within urban areas where there is limited available land for conventional storage on surface. A recently completed prominent underground cavern project is the Jurong rock caverns in Singapore that were the first large caverns constructed in Southeast Asia at a depth of 130 m for the storage of nearly 1.5 million cubic meters of liquid hydrocarbons. Caverns in rock may vary in size from widths of 15 m to 30 m with highly variable heights and lengths.

The excavation of caverns in rock are typically carried out using drill and blast methods in a sequential approach. Firstly, a pilot tunnel, typically of size 4.0 m by 4.0 m or slightly larger, is excavated along the top central position of the cavern to confirm the rock conditions as part of the overall rock support design for the cavern but also to provide the initial access for subsequent excavation stages. Secondly, the remaining haunches or sides of the top portion of the cavern are excavated to expose the entire roof of the cavern and allow for the installation of further rock support. The excavation of the pilot tunnel and haunches is performed using standard drill and blast methods with a drilling jumbo. The remaining volume of the cavern is then excavated in a series of benches, typically 3.0 m to 4.0 m in height, advancing downwards to the final floor elevation of the cavern. Drilling of the benches is usually performed using multiple crawler drilling machines with closely spaced blastholes along the final cavern walls for smooth-wall blasting in order to limit blasting damage of the final walls. In order to facilitate mucking of blast rock a glory hole, typically of 3.0 m size, is excavated from the top bench down to the final floor elevation prior to the excavation of the first bench. Specialized long rock support may be required to be installed into the cavern roof and is commonly only performed after excavation of the first bench in order to have sufficient clearance space. Figure 8.13 presents the excavation of the top section of the underground powerhouse cavern at the John Hart Hydropower project in western Canada.

Figure 8.13 Cavern excavation for hydropower powerhouse.

8.19 New and developing technologies for excavation in rock

Some important innovative advances in excavation technologies for the construction of tunnels in rock have been developed from well-known international equipment suppliers in the tunneling industry.

The Vertical Shaft Machine (VSM) has been developed by Herrenknecht for the excavation of shafts in rock with strength less than 80 MPa that includes a shaft boring machine comprising a telescopic rotating cutting roadheader drum equipped capable to swivel up and down or rotate to excavate the entire cross section of the shaft, and the lowering units. The shaft boring machine is lowered into the launch shaft structure and attached firmly to the shaft with its three machine arms. The lowering units comprise multiple concrete/steel rings with a base ring and cutting edge of the entire shaft size that provide immediate support to the excavated shaft profile and advance downwards vertically facilitated by bentonite lubrication of annular gap in a controlled manner.

Concurrent working activities (excavation, removal of excavated material, shaft construction, and lowering of the shaft structure) make it possible for the VSM to achieve high advance rates of up to 5 meters per shift. The excavated material is removed hydraulically through a submersible pump and transported to the separation plant on the surface. The VSM has been used successfully to date on a limited number of tunnel projects to provide early access to commence tunnel construction.

Small diameter TBMs for rock have been developed and used successfully by the Robbins Company for several tunnels in rock. These small size TBMs are referred to as Small Bore Units (SBUs) and the size of these small TBMs generally ranges 0.6 m to 1.8 m and are capable to excavate rock up to 175 MPa to maximum lengths of 150 m.

Slightly larger sizes of 1.2 m to 2.0 m can be equipped with a motorized head and manual operation and are capable to excavate rock up to 175 MPa to maximum lengths of 300 m. The application of these small size TBMs is becoming increasing interesting to consider for possible use for small conveyance and hydropower tunnels where only a minimum size tunnel is required and the economics of the project is very sensitive to the tunnel construction cost.

The applicability of the use of SBUs should be carefully evaluated for a proposed small size tunnel in rock especially in terms of rock strength and length which are the critical limitations of this technology. A review of similar small size tunnels in rock in similar geology completed using this technology should also be performed. Further developments in this technology are expected to allow for the construction of small size tunnels in rock of great length to benefit the economics of tunnel projects.

The use of small size or microtunnel boring machines (MTBMs) for lake taps for hydropower projects needs to recognize the risk of high pressure groundwater inflows along the tunnel prior to piercing of the lake due to open fractures that are common. For such applications it is advised to only consider the use of a closed face MTBM in conjunction with a fully steel lined or cased tunnel. Steel bulkheads are included in the steel casing behind the MTBM to provide the necessary protection against high pressure leakage through the MTBM as well as the use of appropriate pressure rated seals at the entry portal seal. While many lake taps for hydropower projects only require a minimum hydraulic size, it is typically necessary to oversize the MTBM in order to allow for high thrust disc cutters to work efficiently for practical production and completion of the lake tap tunnel.

A dual-mode or hybrid type of TBM has been developed by the Robbins Company referred to as a "Crossover" TBM which allows for the excavation of mixed ground conditions that might otherwise require multiple tunneling machines. The Crossover TBM includes design features of Single Shield Hard Rock machines and Earth Pressure Balance (EPB) TBMs for efficient excavation in mixed soils with rock. The Crossover TBM has been used successfully for several projects with variable ground conditions and can be expected to be utilized for an increasing number of future tunnel projects as more projects worldwide are planned in difficult and varying ground conditions.

Finally, the application of horizontal directional drilling (HDD) techniques are advancing rapidly to allow for the excavation of small diameter (600 mm) by upward inclined drilling over great distances as much as 500 m through moderately to strong rock (Schmach & Peters, 2016). These advances hold great promise for the future development of mini-hydropower schemes that were previously uneconomic due to the high costs of tunneling.

Further innovative advances in excavation technologies for the construction of tunnels in rock can be expected to be developed in the near future by international equipment suppliers given the increasing demands in the tunneling industry for more economical and rapid excavation techniques.

8.20 Construction methodology evaluation and risks

The choice of method of excavation should be thoroughly evaluated based on the inferred rock conditions and identified key construction risks. For long tunnels it is appropriate to consider multiple types of excavation in order to reduce excavation risks accordingly.

The main types of method of excavation for tunnels in rock are drill and blast and TBM while roadheader excavation is generally limited to short tunnels in low to moderate strength rock.

The optimal method of excavation between drill and blast and TBM options for a tunnel in rock should be evaluated in relation to strong consideration of the following factors:

- Experience of Tunnel Constructor;
- Availability of tunnel equipment;
- Site Access Constraints;
- Tunnel Length;
- Total amount and distribution of poor quality rock conditions;
- Total amount and distribution of potential excessive groundwater inflows;
- Total amount and distribution of low strength rock conditions subject to over-stressing, and;
- Total amount and distribution of geological faults.

Some tunnel constructors in some countries may have limited experience with the use of TBMs while having extensive experience with drill and blast methods based on local historical practice. In the absence of international teaming there will therefore be a bias to adopt drill and blast practice. In some such similar practices, clients without experience or knowledge of the use of TBMs will also prefer to see the use of drill and blast methods.

It is well established practice that the procurement of TBMs requires a longer lead time of a minimum of about 9 months in comparison to drilling jumbos of a minimum of about 6 months. The availability of used TBMs is however significantly reducing procurement times and allowing an increasing number of long tunnel projects to consider the use of used TBMs without start up delays.

While TBMs have been used to date at some very remote mountainous project sites around the world, their mobilization generally requires road or boat access and is limited via air access due to lifting limitations. The largest size of a TBM that is believed that can be transported by helicopter into a remote project site is about 4.0 m. A thorough evaluation of site access and mobilization requirements should be performed as part of the early planning and design process.

Generally, the use of TBMs is cost-effective for tunnels in rock of lengths of a minimum of about 4 km. However, the use of used TBMs may be cost-effective for shorter tunnels. The availability of used TBMs should be thoroughly evaluated as part of the planning and design process to fully recognize the unique benefits of used TBMs that may be applicable.

A proposed tunnel alignment where the tunneling conditions have been characterized to be represented by a large proportion of either poor quality, low strength, excessive groundwater inflows, and/or numerous major geological faults, is deemed to be of high risk and can be expected to result in excessive over-excavation, grouting requirements, and rock support leading to a longer construction schedule and an overall higher cost tunnel project. These characterized tunneling conditions are considered to be of high risk for the application of TBM excavation versus drill and blast excavation.

While the use of TBMs has been proven to be successful among tunnel projects with a significant amount of poor tunneling conditions localized at discrete major geological fault zones, a threshold amount of 30% poor tunneling conditions is considered prudent to warrant the use of drill and blast methods only. This will limit the total amount of delays associated with TBM entrapment at multiple geological faults or within long sections of poor tunneling conditions. Where several major geological faults are present but among a large proportion of good tunneling conditions, it is considered prudent to utilize TBM excavation with the expectation of possible significant delays at the intersection of such geological faults, but the overall construction schedule can be expected to be less than for tunnel excavation using only drill and blast methods.

The highest risk of the use of TBM excavation where several major geological faults are present is the entrapment of the TBM. Based on historical project information, the typical period of entrapment of a TBM at a major geological fault is about 6 months and a bypass tunnel is commonly required to free the TBM and allow continued excavation.

It may be appropriate to utilize multiple types of excavation methods along a tunnel alignment where there are significantly different rock conditions associated with different levels of risk.

Chapter 9

Tunnel support

9.1 General design principles

The support of tunnels in rock is a fundamental design requirement to be implementing during construction in order to stabilize the excavation for the safety of workers and maintain the tunnel profile to facilitate ongoing excavation and completion of construction which may include the installation of final lining.

The support of tunnels in rock is commonly specified as part of construction contract documents and shown on design drawings for construction to present the geometrical layout of the requirements. The type, diameter, capacity, length, spacing and maximum distance from the tunnel face of support components required to be installed are typically presented for design purposes.

The design of support components for tunnels in rock should be based on good industry practice to achieve adequate factors of safety against failure to maintain stability for the designated design life and for the safety of workers at all times during construction. The designated design life should be evaluated and confirmed during the early stages of the design and be based on the purpose of the tunnel for its intended use and the nature of operations.

The design and implementation of support for tunnels in rock should be based on a flexible design approach that can be changed at any time during construction based on the encountered conditions that may be different from those conditions anticipated prior to construction.

The terms "temporary" and "permanent" support were used in historical practice of tunneling in rock but resulted in several contract claims since all support that is installed is not typically removed, but rather left in place as part of the final works for the project. In current practice there is limited cases where the use tunnel support can be regarded as temporary except for typically small temporary works that are completed to allow for the final works to be completed. This work may comprise the construction and securing of temporary working platforms or portals that will be backfilled. For these cases where the duration of use is limited to less than the entire tunnel construction period it is appropriate to consider a temporary design criteria.

All excavations created as part of the construction of a tunnel including niches or chambers for re-mucking, transformers, sumps, refuge bays/stations, ventilation fans, and maintenance shops and offices should be designed with support for the entire tunnel construction period. In many cases, all excavations created for construction will

remain for the operations of the tunnel, and therefore should be designed with support for the long term operational life of the tunnel.

The use of the terms "temporary" and "permanent" support is now discouraged and should be replaced with "initial" and "final" support. Tunnel support can be recognized as being temporary in nature where fiberglass bolts are installed to pre-stabilize an area for raisebore shaft excavation or for the pre-stabilization of the tunnel face for weak and/or highly fractured rock conditions.

The analyses of tunnel stability and tunnel design should be based on achieving a level of safety defined by Factors of Safety. For the design of short term excavations less than the tunnel construction period it is appropriate to base the support on a Factor of Safety of 1.5.

For the design of all long term excavations with use for a duration similar to the tunnel construction period and for all excavations that will remain as part of the tunnel for future operations a minimum Factor of Safety of 2.0 should be adopted for the design of all support.

The design of tunnel support should be presented in a series of design drawings to be issued for bidding and eventually issued for construction. The design of tunnel support for tunnels in rock should include multiple "classes" (Class 1, 2, 3) or "types" (Type 1, 2, 3) to cater for the anticipated total variation in rock conditions along the entire tunnel alignment and defined by increasing levels of support. The minimum level of tunnel support defined as Class 1 or Type 1 should consider the increasing degree of safety commonly demanded in the industry by clients, safety authorities, as well as by the corporate safety policies of tunnel constructors. In general, the minimum level of tunnel support should comprise pattern rock bolts for tunnels with a width greater than 4 m. The use of "spot" bolting at discrete locations typically identified during construction by an engineering geologist represents a high risk design approach and should not of be used due to the risk of misidentifying potentially unstable rock wedges of significant size and the incorrect installation location of rock bolts. Tunnels with a width greater than 4 m with moderately fractured rock conditions commonly result in the formation of potentially unstable wedges of a minimize size of 0.5 m^3, If left unsupported they pose a high risk for the safety of workers and equipment, and therefore warrants a minimum level of rock support comprised of pattern rock bolts. In addition, the minimum level of tunnel support should include pattern rock bolts extended along the tunnel sidewalls below the springline to 1.5 m above the tunnel invert such that high sidewalls are not unsupported.

9.2 Initial rock support

Initial rock support is the portion of the total rock support that is installed immediately after excavation to prevent instability from occurring such as the dislodgement of rock wedges from the tunnel profile. Initial rock support should only be installed after the completion of an appropriate amount of rock scaling and the removal of loose rock fragments present after excavation. Initial rock support typically comprises the installation of rock bolts in conjunction with welded wire mesh, lattice girders, and/or shotcrete. The minimum length of rock bolts comprising the main form of initial rock support should be half the tunnel width for fair to good quality rock conditions. The final design length of all rock bolts should be based on consideration of the identified maximum size potentially unstable wedges that may form around the tunnel profile as well as practical installation requirements.

The drilling of holes for the installation of rock bolts can give rise to the loosening and instability of rock wedges and so care should be taken by all workers by positioning themselves not directly below the installation location. Where highly fractured rock conditions exist it may be appropriate to apply an initial layer of shotcrete immediately after scaling which is commonly referred to as "splash-coat" to provide a degree of safety for workers and equipment before the installation of rock bolts.

All initial support should be installed to within a maximum of 0.5 m behind the tunnel face with mesh and/or shotcrete installed to the corner of the tunnel face and roof and sidewalls. Initial support utilizing rock bolts should be extended over the tunnel profile to a minimum of the springline on each side. For tunnels with high sidewalls all initial support using rock bolts and mesh should extend along the sidewalls to within 1.5 m above the invert to prevent the fall out of rock slabs.

9.3 Final rock support

Final rock support is the portion of the total rock support that is typically installed at a delayed time well after excavation and at a distance from the tunnel face. Final rock support is installed after an evaluation of the behaviour of excavation stability has been undertaken and where it is concluded that additional support is warranted due to a residual risk of instability such as ongoing deformation or loosened rock wedges that is unacceptable. Final rock support typically includes the installation of additional rock bolts and shotcrete.

The design of final rock support for hydraulic tunnels should be based on consideration of any transient pressures that may occur during operations. The requirements of final rock support should not be confused with final lining requirements for hydraulic tunnels. Final rock support should address the overall tunnel stability of those sections that are not subject to scour and/or erosion whereas final lining should address only those sections that are subject to scour and/or erosion during operations. In some cases, additional rock bolts and shotcrete may be warranted to those sections that are not subject to scour and/or erosion but where loosening of the rock conditions has occurred as observed during tunnel construction.

9.4 Practical installation

The design of all rock support systems should strongly consider the minimum excavation dimensions that have to be completed in order to create adequate clearance for the positioning of equipment including drilling jumbos, drilling bolter machines, and hand drilling methods, to allow for the practical installation of rock support.

In some cases it is appropriate to develop a rock support design that is installed in sequence with multiple stages of excavation and allows for the installation of short rock support followed by longer rock support, for large size excavations, only after the required enlargement has been completed. If however the stability of the initial excavation stages requires the installation of long rock support then an alternative approach needs to be considered such as the design of a dedicated rock support gallery to allow for pre-installation of long support. Rock support galleries, excavated above the main excavation, have been used successfully for the pre-installation of long rock support in the roof areas of large caverns site in weak geological conditions.

9.5 Portal support

Tunnel portals represent a key area of construction for the excavation of a tunnel where all workers, equipment, and offices and shops will be present and traverse in and out from the tunnel during construction and therefore is a work area of high risk that requires to be stabilized and maintained during the entire duration of tunnel construction.

Accordingly, the support of tunnel portals is critically important and should be thoroughly evaluated and completed prior to any excavation of the proposed tunnel. A comprehensive inspection and mapping of the tunnel portal areas should be performed as part of the site investigation program to identify any possible hazards such as potentially unstable large rock blocks, unstable trees, soil deposits, surface streams, and groundwater seepage through fracture zones. Portal support designs should be developed to address all identified hazards and consider the final design of the portal. All sources of surface and groundwater should be diverted away from immediately above the tunnel entrance and channeled through a pipeline to prevent scour and/or erosion that may result in loosening of potentially unstable areas.

Tunnel support for portals can include the following types:

- Short length rock bolts;
- Extended length rock bolts or anchors and cables (spiling);
- Canopy tubes;
- Shotcrete encased steel arch ribs/lattice girders;
- Slope mesh;
- Rockfall catch fences;
- Benches and Berms;
- Shotcrete;
- Grass seeding, and;
- Structural canopies.

Figure 9.1 presents portal support works comprising several 8 m long high capacity rock bolts for the stabilization of a large wedge immediately above a tunnel portal.

Perimeter drains should be established at the crest of all tunnel portals to divert all water around the portal area that may be subject to surface run-off from precipitation or natural water sources. All vegetation that poses a potential hazard due to high winds during storm events should be removed in accordance to local environmental and safety regulations.

The stability and excavation of any overlying overburden materials should be evaluated and appropriate designs developed to prevent raveling and sloughing.

Geotechnical instrumentation should be established and monitoring performed on a routine basis during the early stages of excavation for all portals subject to potential unstable hazards. A common form of geotechnical instrumentation includes surface mounted survey prisms drilled into rock. All monitoring results should be immediately evaluated after the completion of surveys and any concerns including minor movements of prisms documented and communicated to the portal and tunnel work crews. In the event of uncertainty associated with the results that indicate possible movement, all portal and tunnel work should be stopped and all workers removed from the work areas, and an evaluation of the stability performed and confirmed prior to the re-commencement of portal and tunnel works.

Figure 9.1 Installation of long rock bolts at tunnel portal for stabilization of large wedge.

9.6 Support components and typical products

9.6.1 Rock bolts

The most common form of tunnel support for tunnels in rock comprises rock bolts fabricated from mild to high grade solid core steel of various diameters typically ranging from 20 mm to 32 mm. The common grade of steel available from many suppliers is 408 MPa (60 ksi) and an enhanced steel grade of 510 MPa (75 ksi) is becoming quite common.

Various types of rock bolts are available from international suppliers and include the following:

* Friction type (Split-set);
* Mechanical end anchored (expansion shell);
* Expansion type;
* Solid core threaded and deformed rebar;
* Hollow-core;

Hoek and Brown (1980) provide descriptions of the various types of rock bolts commonly used in the tunneling industry. Rock bolts are secured inside the drilled holes subject to their type. Friction type are secured based on reduction of the diameter of the bar, mechanical types by a manually set expansion shell at the end, expansion type by air or water pressurization, and rebar and hollow-core type by pre-insertion of pre-fabricated resin cartridges or post injection grouting methods. The highest load-deformation behaviour is realized by solid core steel rebar.

CT Bolts have become a common rock bolt product used for initial support given its design to allow initial securement by a mechanical expansion shell and subsequent full column grouting by pre-placed grouting tubes. The effectiveness of the mechanical expansion shells included with CT bolts and other types of bolts should be routinely evaluated, and if necessary, re-tightened before grouting, when used as part of drill and blast methods after each blast due to potentially loosening.

Enhanced corrosion protection has been developed by several suppliers by means of various types of chemical coatings and plastic covers. Many suppliers also offer rock bolts made from stainless steel which provides enhanced corrosion protection but importantly not for highly acidic groundwater conditions.

An equivalent design standard for the risk of corrosion can be achieved by increasing the diameter of rock bolts and assuming a specific rate of corrosion over the life of the tunnel.

Where there exists a risk of severe corrosion the only design solution for rock support comprises fiberglass rock bolts. Several suppliers fabricate both solid and hollow core fiberglass rock bolts of variable capacity. The main disadvantage of fiberglass rock bolts is their limited shear strength which should be evaluated as part of any design. Fibreglass rock bolts are also commonly used for the support of the roofs of raisebored chambers as well as for pre-support of a tunnel face to provide temporary stability for safe subsequent excavation.

All rock bolts should be nominally tensioned by hand as part of the installation procedure. Quality control testing of all rock bolts should be performed by pull-out tests of at least 10% of the total quantity installed.

The type of rock bolts designed should be evaluated in terms of their acceptability for rapid installation in order to facilitate good excavation production and to meet the long term design requirements. CT and Swellex bolts have been proven to result in good production due to their quick installation. In addition, threaded rebar grouted using resin cartridges has also resulted in good production. While there have been limited historical cases of poor experiences of the use of resin cartridges believed to be associated with an exceedance of the expiry date, the application of resin cartridges as part of a long term rock support system represents a sound and acceptable engineering design.

9.6.2 Cables

Cables or cable bolts are typically used in large width excavations where long and deep support is commonly required to achieve stability. Cable bolts are commonly fabricated as multi-strand cables characterized with high capacity. Cable bolts can be installed with minimum clearance in narrow height excavations prior to enlargement such as for large chambers and caverns.

No form of corrosion protection has yet been developed for cable bolts and therefore they should not be used in potentially corrosive conditions.

9.6.3 Mesh

Welded wire mesh is used for tunnel support for the containment of small rock fragments between medium to large size rock wedges support by rock bolts. The installation of welded wire mesh introduces extra work activity in the total cycle and therefore results in slower rates of advance. Welded wire mesh should be extended over the tunnel profile to the springline at a minimum and extend downwards along the sidewalls to a maximum of 1.5 m above the tunnel invert.

For tunnels in rock where total shotcrete lining is not required it has become common practice that many tunnel constructors require the installation of welded wire mesh over the entire tunnel profile as part of enhanced safety practice.

Welded wire mesh should only be used where shotcreting of designated areas of a tunnel in rock is envisaged to be required as part of the initial rock support system or for scour or erosion protection for hydraulic tunnels.

Chain link mesh may be appropriate as an alternative to welded wire mesh where shotcreting is not expected to be required. The key disadvantage of chain link mesh is that sagging commonly occurs allowing for the collection of loose rock fragment that will be required to be cut and emptied. A high density of mesh pins should be used for the effective installation of chain link mesh to limit sagging. Shotcreting of chain link mesh should not be considered due to the risk of sagging.

9.6.4 Shotcrete

Shotcrete is a very effective method of support for tunnels in rock. Shotcrete can be applied as dry mix with water added at the application nozzle or as an often pre-made wet mix. Plain shotcrete is commonly used in conjunction with welded wire mesh. Fibre reinforced shotcrete provides enhanced compressive and flexural strength and offers the advantage of not having to install welded wire mesh which results in an overall shorter cycle time and higher rates of excavation advance. For these reasons, fibre reinforced shotcrete is increasingly being used in the underground industry. Zhang and Morgan (2015) outline important aspects of a quality control program for the effective use of wet mix fibre reinforced shotcrete.

The early strength behaviour of shotcrete is important to confirm for re-entry into the tunnel face area and commencing the next cycle of activity. Traditional early strength testing of shotcrete has been based on achieving a minimum uniaxial compressive strength of about 4 MPa in 4 hours before re-entry for infrastructure tunnels and 1 MPa for mining tunnels. Saw *et al.* (2015) proposes the use of a shear strength criteria as measured on shotcrete paste using a vane shear apparatus for consideration. Saw *et al.* (2015) further reports that an average layer of about 50 mm thickness develops sufficient shear strength of about 20 kPa within about one hour to support a tetrahedral block with 1 m edge lengths and similarly develops to about 100 kPa in 4 hours to support a cubic meter block and the shotcrete layer. Longer term shotcrete strengths including 3-day, 7-day and 28-day should also be confirmed both based on trial mixes before construction as well as part of a routine testing program during construction. Pre-construction testing of cores drilled from panels or in situ cores from a mock-up test area at a tunnel portal or at a designated shotcrete laboratory or testing site should be completed at least 28 days in advance of any shotcrete applications.

The minimum thickness of shotcrete should not be less than 50 mm for tunnels less than 4 m in width and the minimum thickness of shotcrete should not be less than 75 mm for tunnels greater than 4 m in width.

Project specific certification of all shotcrete nozzlemen should be confirmed prior to all shotcrete application.

Routine testing of cores drilled from panels should be performed during construction at a frequency of three cores per maximum 100 m² of application. In situ cores should be tested at frequency of three cores per maximum 400 m² of application.

Blasting should not be conducted in close proximity to shotcrete of age less than 12 hours.

Technological advances in shotcrete continue to be introduced in the underground industry. A shotcrete specialist should be engaged to confirm the most current applicable testing procedures and specifications as well as provide advisory and testing services for pre-mix designs and nozzlemen certification.

9.6.5 Lattice girders and steel sets

Lattice girders and steel sets are passive types of support that are required for very poor quality rock conditions. A distinct advantage of lattice girders is that they can be easily installed manually without the need for large equipment and are flexible to fit to the tunnel profile geometry. Lattice girders are exclusively used in conjunction with shotcrete and offer enhanced support by acting as reinforcement to shotcrete and are very amenable for use with shotcrete whereby shadowing of shotcrete is very limited. Lattice girders are typically designed and installed at spacings ranging from 0.5 m to 1.0 m and should not be used at spacings greater than 1.5 m. It is however important to recognize that the use of lattice girders for tunnel support will only be effective as the strength of the shotcrete develops and therefore does not serve for immediate or early support. Lattice girders are available as Pantex lattice girders distributed by Dywidag-Systems International (DSI) as standard 3-bar girders for typical tunnel profile support or as 4-bar wallplate beam or stiff cross girders.

Steel sets are preferred and used by some tunnel constructors but are much more time consuming for installation and are at risk of shotcrete shadowing. Steel sets are also typically installed at spacings ranging from 0.5 m to 1.0 m and should not be used at spacings greater than 1.5 m. Steel sets are available as rolled steel "I", "H" or wide flanged (WF) beams that are fabricated to suit the design geometry of the tunnel and are connected horizontally during installation with spacer or spider bars at multiple locations along the beam. It is often useful to include the installation of at least two rock bolts along the lower sidewalls and drilled through or connected to the steel set to provide an integrated installation and pinning of the steel sets along the walls particularly for moderately to highly fractured rock conditions. If completely weathered or very poor quality conditions are present such as clay gouge associated a geological fault the inclusion of rock bolts along the sidewalls will not be effective and are therefore not required.

The design for the use of lattice girders and steel sets for tunnel support should be based on technical analyses and is best completed utilizing numerical modeling techniques by analyzing the imparted loading conditions from representative geotechnical conditions and comparison to the structural capacity of the selected support components in terms of allowable bending moments and shear forces (Hoek et al., 2008). The design of steel sets for relatively small tunnels at shallow depth may be based on the rock loading criterion of Terzaghi (1946) that assumes a weight of broken rock above the tunnel in relation to the geometry of the tunnel. It should however be recognized that the design of tunnel support using steel sets for poor rock conditions associated with geological fault zones should be based on loading conditions of limited vertical extent due to the typical limited width or intersection length of most geological faults that results in an arching effect and decreased loading to the steel sets.

A useful rule-of-thumb design criterion for the use of steel sets for emergency situations at site is that the width of the steel set measured in inches is valid for the same size of the tunnel width measured in meters (ie. 4"=100 mm for 4 m tunnel, 6"=150 mm for 6 m tunnel). Horizontal spider bars spaced evenly around the entire tunnel profile should be used to connect each steel set with each other to maintain the alignment of the support system.

9.7 Tunnel support for severe overstressing and rockbursts

The design of tunnel support for severe overstressing and rockbursting conditions associated with the failure of brittle rock should be based on consideration of the safety of tunnel workers. Tunnel workers are at risk of the dangerous conditions associated with severe overstressing and rockbursting during tunnel construction. For drill and blast excavated tunnels most of the tunnel workers are completely exposed and not protected by equipment due to the nature of the work to install mesh against the tunnel profile, rock bolts, ribs and shotcrete. For TBM excavated tunnels, the cutterhead fingershield offers limited protection of some of the tunnel workers.

The philosophy of tunnel support design for severe overstressing and rockbursting conditions is based on practice from deep mining operations where deformable yielding tunnel support components are utilized to allow for controlled energy absorption with deformation of the rock mass and the prevention of a sudden release of energy. Deformable or yielding tunnel support components include the D-bolt, cone bolt, partially grouted cable bolts, and various types of frictional rock bolts. The behaviour and effectiveness of deformable rock bolts is an ongoing subject of research in the mining industry where high stress conditions are normally encountered. Standard grouted rebar rock bolts should be avoided for the use of tunnel support for severe overstressing and rockbursting conditions due to the occurrence of sudden failure and dislodgement. Figure 9.2 Illustrates a protection canopy used for the safe installation of rock bolts under high stress conditions at the El Platanal hydropower tunnel in Peru. A similar protection canopy could be adopted for the installation of mesh, ribs and shotcrete however it is believed that the work procedures could be significantly impeded for good production. Alternative tunnel support designs should be considered in the presence of such conditions including the use of fiber reinforced shotcrete to provide an initial protection layer and remove the need for tunnel workers to work in close proximity of the tunnel profile.

Figure 9.3 presents the McNally Tunnel Roof Support System™. This tunnel roof support system has been successfully used for severe overstressing and the protection of workers for deep TBM excavated tunnels such as that experienced over an extended length of the tunnel alignment at the Olmos Water Supply Project in Peru.

Figure 9.2 Protection canopy for rock bolt installation.

Figure 9.3 McNally tunnel roof support system™ in TBM tunnel.

9.8 Tunnel support for squeezing conditions

The design of tunnel support for squeezing conditions associated with the failure of ductile rock should be based on consideration of the control of very large deformations that may lead to large scale collapse and the prevention of re-profiling. Squeezing conditions do not pose as great of a risk to tunnel workers in comparison to severe overstressing and rockbursting of brittle rock since the failure process is slow and does not result in the sudden release of rock.

Tunnel support for squeezing conditions has developed in recent times based on a limited number of tunnel projects constructed in very weak rock conditions in which yielding forms of passive tunnel support elements were utilized. One of the early and most successful types of yielding tunnel support for squeezing conditions has been deformable or sliding steel ribs with the Toussaint-Heinztmann (TH) steel ribs. TH steel ribs were successfully utilized for a high deformation zone of the Gotthard Base Rail Tunnel in Switzerland as shown in Figure 9.4 that allowed for 70 cm of radial deformation without the need for re-profiling of the tunnel.

While shotcrete is commonly used for tunnel support, it has a very low deformation capacity and loses its support ability upon failure when subjected to significant squeezing. In order to utilize shotcrete, different types of non-re-useable (sacrificial) yielding elements, or lining stress controllers (LSC), comprising steel cylinders have been developed to be installed within shotcrete windows or pre-formed slots around the tunnel

Figure 9.4 Yielding steel ribs for squeezing ground.

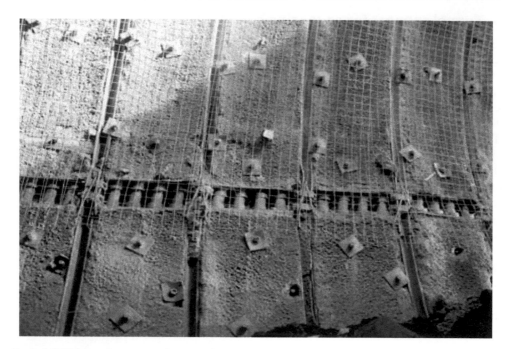

Figure 9.5 Yielding elements in shotcrete lining.

profile. These yielding elements are designed to be loaded axially and deform in stages. Figure 9.5 Illustrates an example of yielding elements in a shotcrete support tunnel which have now become standard practice for the construction of tunnels located in squeezing ground. Larger scale yielding elements have been recently developed and referred to as hiDCon tunnel support yielding elements and used for large size access tunnels as well as for swelling rock conditions, (Barla & Barla, 2009, Kovari, 2012).

Where the tunnel face is also subjected to squeezing conditions it may be appropriate to install fibreglass dowels that serve to limit the amount of deformation and can be easily excavated by a back-excavator.

9.9 Corrosion potential assessment

Rock bolts used for temporary or short term as well as long term design and services requirements may be subject to corrosion during construction if acidic groundwater and/or environmental/ventilation conditions exist. If possible acidic groundwater conditions are present, which is typical for mining projects where the rock conditions are commonly strongly mineralized, groundwater quality testing should be performed as soon as practical from any nearby streams, natural springs or other bodies of water. Groundwater quality information may be available from environmental baseline studies completed during the early stages of the project.

A thorough evaluation of the potential for corrosion should be performed during the early stages of the design process in order to identify this critical risk to long term

Figure 9.6 Softening of shotcrete from direct contact of acidic groundwater.

operations of a tunnel. Multiple methods of assessment are available from Li and Linbald (1999) and the German Standard DIN 50929-3 (1985).

Rock bolts used for hydraulic tunnels will not be subjected to corrosion during operations due to absence of oxygen for fully pressurized flow conditions to allow corrosion. Corrosion may however occur during partial pressure flow operations.

Shotcrete is at risk of deterioration and softening from direct contact of highly acidic groundwater conditions during construction due to ongoing groundwater infiltration. An ongoing evaluation should be performed of the durability and any possible deterioration of shotcrete that has been applied as part of the design for rock support for a tunnel where acidic groundwater infiltration is occurring during excavation. Figure 9.6 presents the results of shotcrete softening and sloughing upon direct contact of highly acidic groundwater inflows with pH of 1.9. Figure 9.7 presents the results of corrosion of a galvanized rock bolt after 60 days due to highly acidic groundwater inflows in direct contact with pH of 1.9.

9.10 Pre-support requirements

Pre-support is the portion of the total rock support that is installed prior to any excavation in order to improve the stability of the rock conditions to prevent instability from occurring upon excavation. Pre-support is typically used for poor quality rock

Figure 9.7 Corrosion of rock bolts from direct contact of acidic groundwater.

conditions where highly fractured and weak rock conditions are present that result in significant geological overbreak. Pre-support is generally installed along the upper part of the tunnel profile along the roof and haunches but may also include installation into the tunnel face.

Typical types of pre-support include rebar rock bolts installed as spiling, hollow-core rock bolts to allow injection and effective integration, fully grouted hollow pipes installed over the partial or full profile of the tunnel as multiple overlapping canopies or umbrellas, and fiberglass dowels installed into the tunnel face. All spiling should comprise high capacity (32 mm diameter) rock bolts of minimum length of one tunnel width, and installed at close spacings of typically 500 mm in order to be effective for tunnel profile control and preventing dislodgment of blocks for moderately fractured rock conditions. In some cases of highly fractured rock conditions, self-drilling hollow-core rock bolts with sacrificial drill bits should be used due to the collapse of the drill holes that prevent subsequent installation. For large size tunnels sited in moderately to highly fractured rock condition with a width greater than 8.0 m only pipe canopies or umbrellas should be considered for pre-support. Pipe canopies or umbrellas are an effective means of pre-support of non-consolidated overburden materials associated with paleochannels should be designed based on consideration of standard industry practice. Typical systems that are available in the industry for pipe canopies include the

Figure 9.8 Pipe umbrella support system.

Alwag AT Pipe Umbrella Support System distributed by DSI and the Robit Casing Forepoling System. Figure 9.8 presents the installation of a pipe umbrella support system in lightly consolidated glacial sediments as part of the construction of the Pinchat double track rail tunnel in Switzerland.

Pre-support typically represents a significant work activity that requires a great amount of time as part of the tunnel construction cycle and will result in slow advance rates.

9.11 Ground freezing

Ground freezing has been a successful form of tunnel support that can be applied either prior to excavation or after to allow excavation of unstable ground and/or cut-off or sealing off of groundwater in highly fractured rock for tunnels subject to high groundwater infiltration from a large local source or sited below the groundwater table or a body of water. Ground freezing has also been used successfully to prevent impacts to groundwater regimes within environmentally sensitive areas.

Ground freezing used for tunnel support must recognize that this method of support is only applied during the period of excavation and must be considered in conjunction with the design and construction of a final lining system that is required to provide the long term stability for the tunnel.

Ground freezing is commonly used for the construction of access shafts through overlying overburden materials and highly fractured, saturated bedrock. Ground freezing should ideally extend into widely fractured or massive rock of low permeability to provide a cut-off of seepage.

The use of ground freezing for tunnel support is a highly specialized subject and a specialized and well experienced ground freezing contractor should be engaged as part of

Figure 9.9 Ground freezing at Hallandsås.

the evaluation and design process. The evaluation of the applicability of using ground freezing for tunnel support should consider a thorough review of historical tunnel projects in similar geological conditions where ground freezing was a proven success.

A most noted example of the recent success of ground freezing was at the Hallandsås Rail Tunnel in Sweden. It required the application of ground freezing in advance of tunnel excavation to prevent unmanageable groundwater inflows and to stabilize poor rock conditions in order to allow for subsequent TBM excavation. Previous attempts at both drill and blast tunnel excavation and shielded TBM excavation in conjunction with standard pre-excavation grouting were unsuccessful. Figure 9.9 illustrates the ground freezing at the Hallandsås Rail Tunnel (Sturk *et al.*, 2011).

Ground freezing has been a successful form of mitigation to adopt during tunnel construction when pre-drainage or grout injection measures have been proven to be too great a challenge or unsuccessful when large groundwater filtration in terms if volumes and pressures have been encountered. As previously discussed, ground freezing can be applied from either within the tunnel or from surface for relatively shallow tunnels. The implementation of ground freezing during tunnel construction should involve a specialized ground freezing constructor with experience in similar conditions to those encountered.

9.12 Tunnel stability and support design verification

Regular inspections should be performed during construction to verify tunnel stability and support design or design modifications that should be made during construction.

While routine inspections should be performed and documented by the technical team of the construction management team, additional inspections should be performed by senior project staff as an independent quality assurance protocol for the project.

Monitoring data from geotechnical instrumentation should be considered as part of the tunnel stability and support design verification process in accordance with an observational design approach. It should however be recognized, that in fair to good quality rock conditions, the stability of an excavation can be rarely presented by using geotechnical instrumentation since rock dilation occurs rapidly, for example with rock block movement or fall out, and is not typically measureable. Geotechnical instrumentation is commonly useful to evaluate tunnel stability for poor to fair quality rock conditions where rock deformation occurs and is measureable.

Chapter 10

Tunnel lining requirements

10.1 Purpose of tunnel linings in rock

Tunnel linings in rock may be required for a variety of reasons including aesthetic, architectural, aerodynamic, hydraulic, operational safety and ventilation, and structural requirements. Tunnel linings in rock may be formed using shotcrete, pre-fabricated segmental concrete, or cast-in-place concrete. In some cases, where good quality rock conditions are present, it may be acceptable that no form of lining is required for long term operations. Common examples of unlined tunnels are for historical and new heavy rail, hydraulic, and traffic tunnels. Shotcrete linings are commonly adopted for tunnels where there is limited access required for humans for inspections during operations such as mine access, conveyor, utilities, and water diversion. Concrete linings are commonly adopted for major infrastructure with extended design lives including drinking water, metro/subway, pedestrian, sewage, storm water, and traffic tunnels.

Tunnel linings in rock have to be designed for all loading conditions for operations.

10.2 Acceptability of unlined tunnels in rock

The evaluation of the acceptability of an unlined tunnel is one of the most important tasks to perform in the early stages of a design process. Unlined tunnels in rock are commonly considered for the conveyance of water either for supply purposes or hydropower generation but may also be considered for mine access and conveyor tunnels as well as traffic tunnels. Numerous unlined tunnels in rock have been constructed for traffic and access in Norway and other parts of Scandinavia. The evaluation must include a thorough assessment of the geological, geotechnical, and hydrogeological conditions along the proposed alignment as well as a full understanding of the operating requirements and any constraints. The evaluation of the acceptability of an unlined tunnel in rock should be updated, and confirmed if possible, during the early stages of construction when the rock conditions are exposed to natural humidity that may have an impact on the long term durability of some types of rocks.

Hydropower projects are typically sensitive to economics and hydropower developers need to understand that the design acceptance of an unlined pressure tunnel represents an economic trade-off between installing a virtually maintenance-free concrete lining versus a partially lined tunnel that requires performing regular inspections and possible maintenance including cleaning of the tunnel and rock trap, which requires easy access to minimize outages and their associated loss of revenue and/or generation for critical infrastructure. Bratveit *et al.* (2016) discuss the recent trends in hydropower energy generation that have resulted in increased demands for peaking production that has resulted in unlined pressures tunnels being subjected to unsteady flow conditions which has triggered increased instability within tunnels. The functional and design requirements for rock traps for unlined pressure tunnels to address such operational challenges are presented by Brox (2016).

The acceptability criteria for unlined water conveyance tunnels includes a large percentage of good to very good quality rock conditions prevailing along the tunnel alignment, low permeability of the host rock mass, absence of adverse or non-durable infilling mineralogy within the host rock types, and limited transient pressures during operations.

In addition, the overall acceptance of an unlined water conveyance tunnel must also importantly be based on the design philosophy that the owner of the project will accept and provide for access facilities in the design and construction of the project. Properly designed access will allow for regular inspections and maintenance as part of the normal operating procedures. Unlined pressure tunnels undergo similar aging as other infrastructure and therefore must be maintained in order to protect the value of the asset. A hydropower tunnel represents a linear structure whereby the failure at a single location will most likely result in a significant impact to the hydraulic conveyance of water. Large failures including partial blockage or full blockage may have a greater impact with the complete termination of operations.

A thorough evaluation should be performed as part of the acceptance of an unlined tunnel and should be based on consideration of the following key issues:

- Operating requirements including safety regulations;
- Long term stability and potential influence to adjacent structures;
- Durability of rock types and fracture infillings particularly subjected to first time saturation;
- Hydrogeological conditions;
- Hydraulic operating conditions with maximum allowable flow velocities;
- Transient pressures, and;
- Provision for easy practical access for inspections and maintenance.

The impact of transient pressures on the stability of unlined tunnels, and in particular, on shotcrete lining of weak seams and non-durable rock, is not well recognized and appreciated. Lang *et al.* (1976) discuss and present examples of the effect of rapid water pressure fluctuations on the stability of hydropower tunnels and provide plausible explanation for the mechanism of the fall out of shotcrete supported weak rock areas. Hedwig (1987) provides an evaluation of the extent of the depressurization and transient pressures along rock fractures within a tunnel wall as part of an overall assessment of the stability of unlined tunnels.

A thorough review and evaluation of transient operating pressures should be performed as part of the acceptability of unlined pressure tunnels.

Figure 10.1 illustrates an example of exposed weak infilled fractures that have deteriorated upon exposure to natural humidity and can be expected to be susceptible to scour and erosion during hydraulic operations.

In the event that a decision is made to design an unlined tunnel, very thorough geological and geotechnical mapping is required to be performed during construction in conjunction with a comprehensive assessment of the final tunnel support

Figure 10.1 Infilled fractures susceptible to scour/erosion.

requirements along all sections of potential adverse geological conditions with full recognition of the loading conditions to be imparted during future operations.

10.3 Shotcrete for final lining

The acceptance of shotcrete for the final lining of a tunnel in rock has been limited in the industry and is subject of the purpose of the tunnel. While shotcrete for final tunnel linings has been generally accepted for tunnels where there is restricted access for personnel such as for mine access, conveyor, utilities, and water diversion tunnels, it remains unaccepted for most major urban infrastructure. A few examples of shotcrete final linings for major urban infrastructure include the London and Stockholm metros, and traffic tunnels in Argentina, Chile, and Norway.

In most cases, shotcrete is not designed as a final lining for long term structural loading conditions of operations but is rather designed and used as initial support during tunnel excavation sometimes referred to as a primary lining inside which is constructed a final concrete lining.

The quality and the quality control of the placement of shotcrete has advanced significantly in the industry with improved mix designs with the use of admixtures, greater training and certification of nozzlemen, and the use of robotic equipment. The key advantage of final shotcrete linings is the flexible application for irregular and varying geometries such as for metro stations and associated platform tunnels as well as for drill and blast excavated tunnels.

The main reason for the lack of general acceptance of shotcrete for more major urban infrastructure tunnels is that shotcrete, unlike, concrete, is more dependent on the quality control of the placement and there exists greater variability and associated uncertainty in the overall quality of the product despite what many shotcrete suppliers and tunnel constructors may perceive. There have been numerous projects whereby the tunnel design consultants have prepared designs for final tunnel linings based on both shotcrete and concrete, and upon review of the similar bid prices, clients have preferred the concrete designs.

The design of a final tunnel lining using shotcrete should be based on an integrated approach whereby the rock support comprising rock bolts is assumed to significantly contribute to the long term stability of the tunnel and as such must be maintained to be effective for the design life of operations. The structural condition and the effectiveness of the rock bolts should therefore be regularly inspected and evaluated during operations and documented as part of the maintenance records of the tunnel.

Further advances in technologies for shotcrete mix designs, quality control and equipment applications can be expected to be developed in the industry which can be expected to result in an increasing acceptance for more tunnel projects.

10.4 Shotcrete for final lining of hydraulic tunnels

Shotcrete is the most common form of tunnel lining for predominantly unlined hydraulic tunnels which are excavated using drill and blast methods.

The main purpose for the use of shotcrete for final lining of hydraulic tunnels is for protection, and not long term structural requirements, to prevent scour and erosion of non-durable rock conditions which typically exists in discrete locations along a tunnel alignment. The minimum thickness of shotcrete used for scour protection in hydraulic tunnels should not be less than 75 mm. Figure 10.2 shows an example of shotcrete final lining for a hydropower tunnel.

Where a significant section of a hydraulic tunnel is associated with non-durable rock conditions it may be more appropriate to place a concrete tunnel lining using telescopic formwork as this may be more economical in comparison to shotcrete.

The placement of shotcrete for scour and erosion protection in hydraulic tunnels should generally extend longitudinally beyond the contact of the poor quality rock conditions for a minimum of 1.5 m in both upstream and downstream directions and also extend around the entire tunnel profile to prevent undercutting of the shotcrete during operations. Where poor quality rock conditions do not appear along the tunnel floor it is necessary to extend the final shotcrete lining to the base of the tunnel sidewalls and place a formed or hand-packed corner infilling to prevent undercutting of the shotcrete along the sidewalls. Where very weak rock conditions are present including incised veins or seams of clay gouge it is necessary to construct seam treatment of the area with the removal of the weak material to a depth of at least 20 cm, followed by an integrated design of mesh, rock bolts, and shotcrete or concrete.

The evaluation of shotcrete final lining for hydraulic tunnels should be performed by a team of engineering geologists to thoroughly identify and map the locations of non-durable rock conditions and in conjunction with tunnel design engineers to confirm the

Figure 10.2 Shotcreting for final lining in a hydropower tunnel.

thickness and areal extents at each location. The design of shotcrete final lining for a hydraulic tunnel should be carefully reviewed by an independent tunnel consultant prior to the commencement of all final lining works.

10.5 Cast-in-place concrete for final lining

Concrete is the most common form of tunnel lining for most types of tunnels in rock for infrastructure. The history of the use of concrete in underground construction and its performance provides confidence for its continued use for future tunnel projects in terms of availability, ease of placement, quality control, and long term durability with minimum maintenance. Cast-in-place concrete linings are commonly designed and constructed as the secondary and final lining for long term loading conditions during operations.

The design of a final tunnel lining using concrete should be based on representative long term loadings contributing from both rock and groundwater pressures. Representative loadings from rock should comprise maximum sized rock wedges based on their formation and location from the intersection of the main sets of rock fractures. Representative loadings from groundwater pressures should comprise a proportion of the maximum theoretical groundwater pressure due to re-establishment of the groundwater regime after tunnel construction and assuming the partial relief of the external groundwater pressures through a drainage system incorporated into the lining system for the design philosophy of a "drained" lining. If the tunnel is to be designed as a watertight structure to prevent any leakage the design needs to be based on an appropriately conservative assumption for the long term external groundwater pressure, which in many cases should assume the original groundwater table which is commonly near surface.

The minimum thickness of cast-in-place concrete linings is generally 300 mm for practical installation with formwork. The thickness of cast-in-place concrete linings can be as much as 500 mm to 750 mm for large long term special loading conditions of operations. Cast-in-place concrete linings of relatively flat or gradual arches along the tunnel roof are required to include steel reinforcement to prevent tension loadings. Steel reinforcement will also be required to be included for cast-in-place concrete linings of significant thickness for shrinkage control purposes. Tunnels that are subjected to or designed for long term swelling pressures due to the presence of swelling types of rocks of clay shales, anhydrite/gypsum, and young volcanics will commonly require a curved structural concrete invert and may also warrant sequential excavation (Steiner *et al.*, 2010, Galera *et al.*, 2014).

The use of concrete for tunnel linings requires the fabrication and placement of formwork. Single specialized formwork may be required for complex geometry whereas travelling and telescopic formwork is available for simple and standard tunnel geometry to allow for high rates of production. Figure 10.3 shows an example of a large diameter traveling collapsible formwork used at a grade of 12.5% for the construction of cast-in-place concrete lining for the tailrace tunnel of the Ingula Pumped Storage Scheme in South Africa. A similar large diameter traveling collapsible formwork was at

Figure 10.3 Large diameter traveling collapsible formwork.

a grade of 17% for the construction of cast-in-place concrete lining for the headrace tunnels at the Waneta Hydropower Project in Canada.

The use of concrete for tunnel linings in drill and blast excavated tunnels requires void filling or contact grouting after the initial placement and setting to provide a complete seal of all voids due to shrinkage. Prior to the placement of formwork for a drill and blast excavated tunnel the locations of all large voids along the tunnel profile should be accurately surveyed so post-grouting sleeves can be included in the formwork or locations for grout holes can be confirmed.

Ongoing advances in concrete technology include pumpability over long distances for long tunnels, fire protection to limit damage, reduced permeability, and early strength attainment to allow early stripping of forms to achieve high rates of production.

10.6 One-pass concrete segmental lining with TBM excavation

The construction of tunnels in rock using TBMs in conjunction with one-pass segmental concrete linings offers a unique and cost-effective design solution when warranted. Numerous tunnels in rock have been successfully completed using TBMs in conjunction with one-pass segmental concrete linings where variable and challenging rock conditions were present. These tunnel projects have included metro, mine access, water supply, and hydropower tunnels.

This specific tunnel construction approach is commonly adopted when the long term durability of the rock conditions is uncertain or suspected of deterioration for the safe and uninterrupted conveyance of water and hydropower generation. The main benefit of this construction approach is the greatly reduced construction schedule as a one-pass construction process in comparison to traditional excavation followed by a secondary activity of cast-in-place concrete lining. Pre-cast concrete segments used for linings in rock tunnels are typically unbolted and non-gasketed and represent a leaky tunnel lining to dissipate groundwater pressures upon dewatering.

One-pass concrete segmental linings have been successfully used for a variety of rock tunnels since the early-1970s ranging from good quality granitic rock in Hong Kong, highly disturbed flysch rock in Greece, non-durable basalts for the Mohale transfer tunnel in Lesotho, poor quality rock conditions at the Inland Feeder Arrowhead Tunnels in the USA, poorly cemented volcanic lahar deposits with high groundwater pressure in Panama, and low strength sedimentary rocks under high cover for the Kishanganga hydropower project in India.

TBM tunnel construction in conjunction with one-pass segmental concrete linings is increasingly being used as the preferred method of construction for hydropower projects that are being developed with major tunnels sited in rock conditions that are deemed to be suspect for uninterrupted long term operations. In addition, there is an increased amount of combined stormwater overflow tunnels that are being planned in urban areas within rock and the preferred design and construction solution is a one-pass segmental lining followed by a secondary inner lining that is typically cast-in-place or pipe backfilled.

The first use of one-pass concrete segmental linings for a hydropower pressure tunnel was for the Orichella Tunnel in Italy in the early 1970s. Since 1990, more than 300 kilometers of one-pass concrete segmental linings have been successfully used for the

Table 10.1 Pre-cast concrete segmentally lined hydropower tunnels.

Project	Country	Year	Length, km	Size, m
Los Rosales	Columbia	1990	9	3.5
Evinos Moros	Greece	1992	30	3.5
Delivery Tunnel North	South Africa	1995	15	5
Cleuson Dixence	Switzerland	1998	7.1	5.8
Manubi	Ecuador	1998	8.3	4.88
Yellow River	China	1999	12.2	6.1
Umiray Angat	Philippines	2000	13	4.9
Val Viola	Italy	2002	18.9	3.7
Doblar	Slovenia	2002	4.0	6.98
Mohale	Lesotho	2002	16	4.9
Plave	Italy	2002	6	6.98
San Francisco	Ecuador	2006	9.7	7.4
La Joya	Costa Rica	2006	7.9	6.2
Talave	Spain	2007	7.5	4.0
Gigel Gibe II	Ethiopia	2008	26	7.0
Beles	Ethiopia	2008	19.2	8.1
Palomina	Dominican Rep.	2008	16.5	4.4
Pando	Panama	2010	9	4.5
Kishanganga	India	2010	14.6	6.1
Suruc	Turkey	2010	17	7.3
Coca Coda Sinclair	Ecuador	2012	24	9.0
Tapovan Vishnugad	India	2012	12	5.6
Kargi	Turkey	2014	11.8	10.0
Xe-Pian Xe-Namnov	Laos	2015	11.8	5.7

construction of hydropower pressure tunnels around the world as presented in Table 10.1. The use of hexagonal pre-cast segmental lining has been preferably used for several hydropower pressure tunnels and has achieved progress rates in excess of 1500 m per month.

A key risk associated with the use of pre-cast segmental linings for tunnels in rock is overstressing of the lining due to highly deformable weak rock under high cover or squeezing conditions. The use of double-shield TBMs have proven to be able to the most reliable and lowest risk approach of TBMs in conjunction with pre-cast concrete segments for the construction of tunnels in weak and/or poor to fair quality rock. The risk of weak rock conditions along a tunnel alignment should be thoroughly evaluated and special lining design measures incorporated such as high capacity concrete or steel lining segments.

Pre-cast concrete segments have been designed with the use of steel fibres as a cost savings design and can provide a similar level of structural capacity in comparison to standard steel bar reinforcement. The use of steel fibres is also being used for pre-cast segmental lining in seismic regions and designed to withstand dynamic loading conditions. The use of steel fibres for the reinforcement of pre-cast segments is preferred in saline groundwater areas to prevent long term corrosion that has occurred in some urban areas including Hong Kong.

Figure 10.4 Pre-cast concrete segmental lining for final lining in a hydropower tunnel.

Most pre-cast segmental concrete linings are bolted and gasketed for use as fully watertight linings for metro tunnels located under the groundwater table. However, non-bolted and non-gasketed hexagonal pre-cast segmental linings have been commonly used for the construction of hydropower tunnel linings where inward or outward leakage is acceptable. Figure 10.4 illustrates an example of hexagonal pre-cast concrete segmental lining.

10.7 Waterproofing requirements and applications

The majority types of tunnels in rock for infrastructure include an operational requirement to prevent water ingress with complete watertightness in order to limit hazards of slipping, ponding, and ice formation. The most common and cost-effective form for the prevention of water ingress in tunnels is the application of waterproofing by use of fabricated polyvinyl chloride (PVC) sheeting membranes or spray-on coatings.

Waterproof membranes are placed against the excavated and support tunnel profile prior to the placement of the final tunnel lining.

Waterproofing membranes may be used in conjunction with a final lining of either shotcrete or concrete. PVC sheeting membranes are commonly used in conjunction with geotextiles/fleeces for drainage and in combination with concrete linings and are

Figure 10.5 PVC sheeting membrane.

attached to the tunnel profile using fabricated connection pins that are heat welded together with the various sheets to prevent leakage. The use of sheeting membranes is well established and is flexible for effective placement for drill and blast excavated tunnels. Sheeting membranes have been used in conjunction with final shotcrete linings and require a sandwich design incorporating either mesh spiders or mats of reinforcing steel within the shotcrete to achieve the necessary bonding between the shotcrete and the membrane. Figure 10.5 illustrates an example of the installation of PVC sheet sheeting membrane in a traffic tunnel.

Spray-on coatings were generally developed for use in conjunction with a final lining of shotcrete. Spray-on coatings are generally required to be applied to an initial layer of shotcrete for effective bonding and are applied using robotic equipment similar to and after shotcreting operations with computerized thickness control. Final shotcrete layers can be subsequently applied over spray-on coatings as part of a sandwich lining design. Figure 10.6 illustrates an example of spray-on waterproof membrane applied over shotcrete tunnel support for a traffic tunnel in the Faroe Islands. Spray-on waterproof membranes require relatively dry surfaces for good adhesion to prevent de-bonding which is typically achieved with well-planned drainage measures.

10.8 Fire protection requirements

The majority types of tunnels in rock for infrastructure are required to include for fire protection of the final lining so there is no risk of a catastrophic tunnel collapse due to a major fire event in the tunnel with the loss of the tunnel lining.

Figure 10.6 Spray-on membrane for waterproofing of road tunnel.

Fire protection of tunnel linings of concrete include the use of polypropylene fibres as part of the mix design for the concrete that limits the extent of spalling under fire temperatures. In comparison, fire retardant spray-on coatings or thermal barriers are available to be used in conjunction with final shotcrete linings or in addition to cast-in-place concrete linings.

Chapter 11

Construction considerations

11.1 Site mobilization

The mobilization of equipment for tunnel construction commonly requires special access and clearance requirements along existing roads, over bridges for wide and heavy loads. A routing study should be performed as part of the early design process to confirm acceptable access to the project site or to identify upgrades, and new access construction that is required. To avoid major upgrades or new construction it may be appropriate to consider alternative transportation such by sea or over other large bodies of water. For remote project sites it may be necessary to consider the use of large payload air transport as either cargo airplanes or helicopters that can be expected to require the disassembly and re-assembly of some large equipment. The mobilization of TBMs is commonly achieved with the transport of the separated components and the use of specialized multi-axle transport bogeys.

The total duration for site mobilization will depend on the availability of existing equipment that is proposed to be used and the procurement time required for new equipment.

11.2 Site preparation of camps, staging and laydown areas

The site preparation of construction camps, staging areas, storage laydowns, spoil disposal sites, and office and shop areas for the designated tunnel constructor can be completed as advanced contracts by the client if the properties of the various sites have been confirmed for use and designs prepared as part of the early design process. Site offices can also be established for the client and consultants prior to the start of construction at designated sites made available by the client. All facility sites provided in advance by the client or designated to be prepared by the tunnel constructor should be thoroughly evaluated for risks during construction including geohazards of floods, avalanches, rockfalls and debris flows in nearby creeks, particularly in seismic regions.

The site preparation of areas for tunnel construction should be consistent with the designs presented on the issued design drawings so that the tunnel constructor fully appreciates what is being provided by the client. All site preparation designs should include adequate drainage and traffic access requirements to avoid delays or impacts during construction. Modifications to the site preparation may be appropriate to consider during the final negotiations with the preferred tunnel constructor or shortly after the award of the contract to tailor any specific requirements of the tunnel

Figure 11.1 Site preparation.

constructor. Figure 11.1 shows an example of advanced site preparation of a tunnel portal.

11.3 Portal and shaft access

The preparation and construction of the tunnel portals or shaft access requirements may also be planned to be completed under advanced contracts if the properties of the various sites have been confirmed for use and designs prepared as part of the early design process.

As part of any early preparation and construction of the tunnel portals or shaft accesses it is not recommended to include any form of construction access such as cranes or mucking facilities so as to allow the tunnel constructor to establish their preferred equipment for construction. All tunnel portal and shaft access designs prepared in advance should include adequate drainage and traffic access requirements to avoid delays or impacts during construction.

11.4 Ventilation

Appropriate ventilation should be established to facilitate tunnel construction by the tunnel constructor in accordance with the local national health and safety regulations. All formal submittals to the safety authorities regarding ventilation should be the responsibility of the tunnel constructor as the extent of the ventilation systems required for construction is generally dependent upon the size of the proposed workforce and type of equipment to be operating underground during construction.

A preliminary assessment of the anticipated construction ventilation requirements should be performed during the early design process. This assessment should confirm that the requirements of the ventilation are in accordance with national health and safety regulations as well as consistent with the proposed tunnel geometry for practical construction.

11.5 Construction water supply

The supply of clean, but non-potable, water for tunnel construction should be evaluated during the early stages of the design process to confirm the availability for use. In some cases, the client may establish the water supply to the site or disclose any access restrictions to local water sources. At remote sites it may be necessary to construct a water well and these requirements, including any environmental regulations, need to be confirmed and issued as part of the design as to which party will be responsible for establishing and maintaining this requirement.

The typical construction water supply requirements for the use of drill and blast drilling jumbos is about 2–3 liters/second and for medium size TBMs is about 3–5 liters/second. Additional water is typically required to be established at project sites during construction for offices and dry/change houses for showers and toilets.

11.6 Electrical supply

The availability of a stable electrical power source and the loads of existing infrastructure as well as the total load requirements anticipated for tunnel construction should be evaluated during the early stages of the design process to confirm the availability for use and the needs for any temporary or permanent upgrades.

For the use of TBMs for tunnel construction the client should establish a connection as part of the site preparation from the local electrical grid to the construction site in order to save significant costs in comparison to the use of diesel generators.

Electrical requirements for drill and blast operations are small and typically less than 2 MW whereas TBM operations, depending on the size of the TBM, can require up to 10 MW.

11.7 Construction pumps and sumps

High capacity construction pumps of adequate capacity should be maintained and utilized during all tunnel excavation that advances at a down grade. The capacity of the pumps should be based on consideration of the anticipated groundwater infiltration. Construction pumps and sumps of adequate capacity should also be maintained and utilized during all tunnel excavation as part of good housekeeping and safety practice in the tunnel.

Flowmeters should be required to be established at multiple locations along the tunnel alignment as the tunnel is advanced if large groundwater infiltration is anticipated and occurs during construction to confirm flow volumes and their possible impact to tunnel construction. A flowmeter or water weir should also be established at the tunnel portals to provide estimates of the total construction water from the tunnel and be recorded on a regular basis during construction.

11.8 Groundwater and construction water treatment

Tunnel construction water is generated as part of the excavation process since it is used for drilling with drilling jumbos as well as with TBMs. The volume or capacity of tunnel construction water is typically limited to less 5 l/s. However, tunnel construction water commonly mixes with groundwater inflows to form the total volume exiting the tunnel during excavation which is required to be treated before release into the environment.

A water treatment facility and total system including settling ponds or tanks needs to be conservatively designed to process the maximum possible construction water and groundwater expected during excavation. Several equipment suppliers provide specially designed water treatment systems for tunnel projects including tanks, piping systems, flocculant products, and pond liners.

Adequate space near the portal is required to operate a multi-stage circuit of sediment ponds extending over an area of typically 10 m wide by up to 50 m long and up to 1.5 m deep to provide adequate settling time. Figure 11.2 presents an example of a

Figure 11.2 Construction water treatment system.

Figure 11.3 Construction water treatment ponds.

construction water treatment system comprising a series of settling containers in operation at the Irvington Tunnel portal. Alternatively, a series of ponds can be excavated if a larger space of land is available near the tunnel portal as shown in Figure 11.3.

Water treatment commonly includes the addition of carbon dioxide (CO_2) to increase the pH level. When construction water comes into contact with shotcrete or when acid rock conditions are present, lime can be added to the construction water to increase the pH level and precipitate metals out of the solution.

11.9 Environmental sampling and testing requirements

Environmental sampling and testing of groundwater and representative rock types are typically performed prior to the construction of a tunnel in rock as part of the environmental studies to evaluate the potential for acidic groundwater conditions and acid rock drainage (ARD) and metal leaching (ML) of rock.

Sampling and testing of groundwater should be performed from boreholes as well as any springs along the tunnel alignment in order to identify the potential for acidic conditions that may have an impact on the design of certain tunnel components.

Similarly, sampling for acid base accounting (ABA) testing of representative rock types should be performed from boreholes as well as any outcrops along the tunnel alignment in order to identify the potential for acid rock conditions that may have an impact on the design of the tunnel and the designated spoil disposal site.

Spoil from tunnels constructed in rock should essentially be characterized during construction as either being potentially acid generating (PAG) or non-acid generating (NAG) in order that the correct disposal can be quickly confirmed without undue interruption and delays to tunnel construction. Rapid and reliable site testing procedures and equipment have been developed that should be established and performed by an independent party during construction with direct reporting to the tunnel constructor but with responsibility to the client.

The recommended procedures to be adopted for ARD and ML sampling and testing is that of Price and Errington (1998) which was developed as a government standard for mine sites in British Columbia, Canada, or the Acid Rock Drainage Prediction Manual (Coastech Research, 2008), which was developed for the Mine Environment Neutral Drainage (MEND) program of the Ministry of Mines, Energy and Resources of Canada. Both documents are widely used and accepted in the international mining industry.

Routine testing of spoil from each blast and/or each TBM stroke advance should be performed during construction. These requirements may necessitate the double handling of spoil with initial dumping at the portal from which sampling can be completed. Final disposal is then commonly confirmed from the testing results.

11.10 Spoil disposal

The disposal of spoil from tunnel construction typically requires the transport of the disposal material that may include fragmented rock from drill and blast operations or small size rock chips from TBM operations. Multiple spoil disposal sites are commonly required to be established for Potentially Acid Generating (PAG) and Non-Acid Generating (NAG) spoil types based on site testing. Figure 11.4 shows an example of a spoil site located adjacent to an intermediate access adit for a long tunnel that has been filled and contoured into the topographic depression of a creek. When an adjacent location near a tunnel portal or shaft is not available for the permanent disposal of spoil it is necessary to provide a tipping station as shown in Figure 11.5 to facilitate the haulage of the spoil to a distant site. Figure 11.6 illustrates the different types of spoil created from different types of tunnel construction including drill and blast, TBM, raisebore drilling, and roadheader.

Spoil from the construction of tunnels in rock may be associated with mineral constituents including sulphides (pyrite) that when disposed of at a designated site and comes into contact with air and precipitation undergoes oxidation, and result in the generation of acidic rock drainage (ARD) and the metal leaching (ML) that may impact the environment. Tunnels proposed for mining projects are commonly associated with ARD and ML. Spoil from tunnels in rock proposed in other locations may also be at risk of ARD and ML and should be evaluated during the early stages of design.

Spoil sites should be thoroughly evaluated for acceptability during the early stages of design in order that environmental baseline sampling and studies can be performed. Spoil sites in close proximity to the tunnel location should be evaluated and established

Figure 11.4 Tunnel spoil site.

if possible to limit the distance of transport. The geotechnical stability of all spoil sites should be thoroughly evaluated with site specific site investigations is necessary with designs developed for long term disposal including drainage and groundwater collection systems if appropriate.

The management and operation of the disposal of spoil is a significant work activity that requires transportation planning and environmental restrictions in accordance with any requirements imposed by the client or local authorities. It is common practice that many clients wish to restrict or prevent the transport of spoil using large trucks along public roads at distances far away from the construction site. For spoil sites located close to the tunnel portals it may be cost effective to utilize a conveyor system for the transport of spoil.

11.11 Tunnel support design implementation

The implementation of a tunnel support design for a proposed tunnel should be based on the original design developed based on stability analyses using data from the geotechnical site investigation program and presented in the format of design drawings issued for construction. The design implementation process should be based on an observational approach whereby the stability of the tunnel is evaluated based on visual inspections and supported with geotechnical instrumentation when warranted, which is typically for weak rock conditions.

Figure 11.5 Tipping station for spoil disposal.

The stability of the tunnel should be re-evaluated during the early stages of construction based on the encountered conditions and modifications to the original design should be undertaken if warranted by the encountered conditions and upon completion of updated stability analyses during construction. Modifications to the original tunnel support design should be presented as revised or new design drawings issued for construction. If the design modifications are significant then the original tunnel support designs should be removed from implementation and distribution so as to avoid any confusion and possible incorrect implementation.

The implementation of the tunnel support design should comprise the evaluation of the encountered rock conditions after each blast advance for drill and blast operations or TBM stroke advance and the formal issuing of a Tunnel Support Instruction (TSI) to document the design instruction between the tunnel designer and the tunnel constructor. The TSI should be distributed to all relevant parties at the project site. For TBM excavation it is acceptable to issue a TSI for a specific section of the tunnel with similar rock conditions which may be applicable to multiple working shifts and only a new TSI when the rock conditions have changed significantly to warrant a new TSI. For weak, moderately to highly fractured rock conditions it is necessary to fully review and evaluate available data and results from geotechnical instrumentation to understand the status of tunnel stability with comparisons to the tunnel support installed and stability of previous sections of the tunnel.

Rock spoil a) drill and blast, b) TBM, c) raisebore, d) roadheader

Figure 11.6 Different types of spoil from rock tunnels.

Routine visits and inspections should be performed by a senior tunnel engineer as a quality control process of the tunnel support design implementation to confirm the timely and completeness of issuing the TSIs as well as the ongoing stability of all sections along the tunnel alignment. The requirements for any additional support further to the original TSI should also be documented by an additional TSI for the designated section of the tunnel.

11.12 Geological and geotechnical mapping requirements

The construction management and resident engineering team should include qualified engineering geologists for routine geological and geotechnical mapping that should be performed on an ongoing basis after each blast for drill and blast excavated tunnels and per shift advance for TBM excavated tunnels during construction.

Standard mapping templates should be established and present all relevant geological and geotechnical information including rock type, degree of weathering or alteration, occurrence, orientations, and nature of rock fracturing, occurrence, orientations, and the nature of geological faults, estimated rock strength, groundwater infiltration, and geological overbreak.

All geological and geotechnical information that is considered to be of particular relevance to tunnel stability, or any observations of instability, should be communicated as soon as possible to the tunnel design team and an updated evaluation of tunnel support performed

Samples of the different types of rock and any fault materials encountered during excavation should be collected and maintained in the site offices for reference. Samples should also be collected of suspect fracture infilling materials and preserved for x-ray diffraction (XRD) testing to evaluate for non-durable mineral constituents.

All geological and geotechnical mapping should be regularly checked by a senior engineering geologist as part of quality control practices.

Finally, inspections for any suspect or actual deterioration and/or instability should be made on a regular basis during tunnel construction and these observations documented including good quality photographs.

11.13 Quality assurance inspections

A senior tunnel engineer who is not involved on a day to day basis during tunnel construction should be engaged on a monthly basis to perform quality assurance inspections on behalf of the tunnel design team. This engineer, often the designated Engineer of Record, should assure the overall stability of all sections of the tunnel and to inspect any possible deterioration or non-performance of rock support. Monthly review meetings should be held between the senior tunnel engineer and the site tunnel team to discuss all findings and any concerns or new observations and information to be documented.

11.14 Geotechnical instrumentation

Geotechnical instrumentation is commonly utilized to provide verification of tunnel stability and support design as well as quality control for controlled drill and blast excavation.

Typical geotechnical instrumentation for rock tunnels includes the following:

- Surface level prisms along the surface of the alignment and at portals;
- Slope indicators along adjacent buildings;
- Tiltmeters along adjacent buildings;
- Tape extensometers;
- Borehole extensometers;
- Rock bolts load cells;
- Shotcrete pressure cells;
- Crackmeters;
- Strain gauges;
- Vibrations, and/or accelerations;
- Hydraulic Pressure Transducers,
- Piezometers, and;
- Flowmeters.

Surface level prisms and tape extensometers are used to measure the movement of rock along the excavation profile whereas borehole extensometers are used to measure the

Figure 11.7 Geotechnical instrumentation – tape extensometer.

movement of rock within the surrounding rock mass around an excavation and tape extensometers are typically used to measure convergence of a tunnel profile. Rock bolt load cells are useful to measure the loads being realized on rock bolts and shotcrete pressure cells are used to measure the loads developed within shotcrete linings. Figure 11.7 illustrates the use of a tape extensometer for the measurement of convergence of the sidewalls of a tunnel.

An instrumentation monitoring program should include the regular collection of data, processing and calibrating of data, plotting of data in relation to excavation progress and geological conditions, interpretation of the results, and action if required. Routine graphical formats of data should be established for all parties to be able to review and understand. The volume of geotechnical instrumentation data can be extensive for many tunnel projects and is best handled through the utilization of specialized software for rapid processing and plotting to allow rapid response when necessary.

Trigger and alarm levels should be developed along with a sound and practical action/communication plan as part of the design implementation in relation to the anticipated deformations. The review and evaluation of instrumentation monitoring data is perhaps one of the most important responsibilities during construction and should not be designated to staff with limited experience.

Blasting vibrations are commonly required to be limited to prevent damage to existing adjacent or overlying infrastructure including the tunnel works under construction as well as to prevent any environmental impacts. Monitoring of blasting vibrations is therefore commonly required to be performed for projects where sensitive infrastructure or wildlife may be present. The most common parameters to be monitored are the peak particle velocity, acceleration, and air overpressure. Multi-channel and parameter seismographs are well established to provide all the monitoring requirements including transfer of data to be reviewed in a timely manner after each blast. Blasting seismographs need to be located and protected in appropriate positions to be effective in recording representative vibration data. For projects where drill and blast excavation is planned and it is perceived that vibrations will not be realized it should be recognized that monitoring should be performed regardless in order to document key information in the event that claims and complaints are made after the works.

The monitoring of the hydraulic operations of water conveyance and hydropower tunnels may be important for some projects where the final design may be considered to be suspect either due to the non-recognition of adverse geological conditions or poor quality of construction. In order to detect any possible pressure decrease that is commonly related to a partial blockage whereby the flow velocity can be expected to increase, it may be appropriate to install hydraulic pressure transducers at practical locations along a tunnel alignment. Practical locations for the installation of pressure transducers include along the inside walls of concrete plugs or bulkheads constructed at intermediate access adits used for tunnel construction. The installation of pressure transducers at multiple locations along a tunnel alignment will allow for the determination of the location of a potential pressure decrease and related problem thus allowing for effective planning for an unwatered ROV inspection or a dewatered manual inspection of the area in question. Multiple hydraulic pressure transducers should ideally be installed at each location for redundancy in case of failure of some instruments during long term operations. Pressure transducers are then connected to a datalogger fixed at the adit portal via cables along the access adit for the monitoring and evaluation of data by the operations team. The frequency of monitoring of hydraulic operations should be monthly during the initial year of operations in order to establish a baseline of information. The frequency of monitoring can be reduced to bi-annually or annually for subsequent future operations.

Chapter 12

Construction risks and mitigation measures

12.1 Portal hazards

Tunnel portals are subject to key risks that may significantly impact tunnel construction including floods by adjacent streams or rivers, landslides from unstable overlying overburden, rockfalls from unstable rock blocks that were not stabilized, and avalanches in cold climate regions where deep snow is present.

Tunnel portals need to be designed accordingly to minimize all risks for the entire duration of tunnel construction and future operations. A thorough evaluation of the prevailing conditions at the designated portal sites should be performed during the early stages of the design to identify all possible risks and mitigation measures should be designed to address all key risks, both that may be present during construction, but also all risks to long terms operations as well, unless alternative portal sites are possible. In some cases it may be appropriate to apply temporary mitigation measures during tunnel construction and develop alternative design solutions for long term tunnel operations.

Figure 12.1 presents a structural canopy constructed for the protection of avalanches at a tunnel portal.

12.2 Tunneling hazards

The construction of tunnels is associated with some of the most severe hazards that may impact successful completion. The most common types of hazards that may occur during tunnel construction are related to the geological conditions and include the following:

- Paleochannels of unconsolidated materials within shallow bedrock;
- Fall out of medium and large size rock wedges;
- Intersection of weak rock associated with geological faults and instability/collapse;
- Stress re-distribution and associated rockbursts upon the intersection of faults;
- Running or flowing ground in conjunction with pressurized groundwater;
- Squeezing of very weak rock conditions;
- Intersection of unmanageable groundwater infiltration;
- Intersection of unmanageable geothermal groundwater infiltration;
- Impact to local groundwater regime;
- Generation of excess fine materials during TBM excavation;

Figure 12.1 Structural portal canopy for avalanche risk.

- Settlement of ground above the tunnel and damage to overlying or adjacent infrastructure;
- Stability influence between adjacent tunnels;
- Stability influence to an existing tunnel;
- Intersection of combustible and dangerous gases including methane (CH_4) and hydrogen sulfide (H_2S);
- Intersection of karstic formations with voids of loose, weak materials;
- Overstressing of weak rock conditions in conjunction with elevated in situ stresses, and;
- Equipment breakdown or severe damage due to abrasive rock conditions;

While mitigation measures can be developed for most of the typical tunneling hazards, such as the implementation of breastboards for the initial support of running ground, it should be recognized that such measures may not entirely prevent impacts to the project schedule and cost, but rather at least reduce such significant impacts. The intersection of geothermal groundwater infiltration may occur in areas along the tectonic plate margins and is commonly associated with dangerous gases including methane and hydrogen sulfide that requires enhanced ventilation measures to be implemented during tunnel construction.

Figure 12.2 presents excavation of a decline access tunnel to an underground hydro-power cavern through an unexpected paleochannel of infilled unconsolidated

Figure 12.2 Canopy tube installation through paleochannel sediments.

sediments that required the installation of multiple umbrellas of drilled and grouted forepoling.

12.3 Stability influence between adjacent and existing tunnels

The excavation of new tunnels in close proximity to each other as well as in close proximity to an existing tunnel is commonly required due to space restrictions for the planning and the alignment design for new tunnel projects.

In general, medium to large size tunnels to be sited within poor to fair quality rock conditions and adjacent to each other or existing tunnels should be carefully evaluated using three-dimensional computer software programs to assess excavation stability and support requirements and provide a basis of predicted anticipated deformations. It is common practice to expect that if appreciable separation distances cannot be adopted, then careful sequencing of excavation will be necessary in conjunction with pre-support of the first advanced or existing tunnel. This approach is necessary in order to limit unacceptable deformations and prevent damage to all tunnels as well as overlying or adjacent structures. In addition, a comprehensive geotechnical instrumentation program is typically implemented to closely monitor the actual deformation and compare to predicted data such that mitigation measures can be adopted in the event that unexpected deformations occur. Typical examples are the pre-support or stiffening of an existing subway tunnel using steel ribs and shotcrete within the existing tunnel profile when a new tunnel is constructed nearby. Similarly, the advancing faces of large size tunnels sited within poor to fair quality rock conditions should be appropriately staggered often by several tunnel widths, particularly within a high in situ stress regime, in order to prevent influence between tunnels.

12.4 Groundwater control and management

The control and management of groundwater infiltration is one of the most important requirements during tunnel construction since inundation of a tunnel can result in significant damage to equipment and also impact the stability of the tunnel and the safety of workers. Groundwater infiltration during tunnel construction may also result in an impact to the local groundwater regime due to the drainage and drawdown effect of tunnel construction. The risk of significant groundwater infiltration should be thoroughly evaluated during the early stages of design and estimates of infiltration presented and included in the risk register and contract documents to convey the severity of the risk to all bidders. The requirement for appropriate mitigation measures and pumps and measuring equipment including flowmeters should be included in the technical specifications in order to be able to manage this risk if it occurs.

An often unrecognized risk associated with rock tunnels is the potential impact to the groundwater regime. In sensitive project locations environmental regulations may preclude any form of impact to the natural groundwater regime where the groundwater regime. This is especially true where the groundwater is relied is relied upon for local groundwater supply or for environmental preservation. Furthermore, areas along a tunnel alignment with relatively shallow overburden overlying bedrock may be susceptible to settlement in the event of uncontrolled groundwater infiltration that can result in drawdown of the groundwater table. This risk is exacerbated in areas where the quality of the shallow bedrock is poor with open fractures of high permeability and/or in association with low in situ stresses.

Mitigation measures to control impacts to the groundwater regime include the utilization of pre-excavation grouting in conjunction with a final permanent non-drained concrete lining. A possible alternative design solution to minimize any possible impact to the groundwater regime is to utilize an earth pressure balanced (EPB) TBM in conjunction with pre-cast concrete lining.

While many designers attempt to address the control of groundwater impacts with technical specifications and a maximum allowable infiltration rate, the implementation of practical mitigation methods with fair payment provisions are the key challenges for tunnel constructors.

Some key examples of the challenges of these risks were realized at the Inland Feeder Arrowhead Tunnels in the USA, and the Hallandsås Rail Tunnel in Sweden. Fulcher et al. (2008) presents the challenges that were experienced with pre-excavation grouting to control groundwater and for ground improvement for the two approximately 6.5 km Arrowhead East and West water transfer tunnels that were subjected to environmental restrictions pertaining to the groundwater table. The initial construction of the Arrowhead West tunnel resulted in cumulative groundwater inflows of 5500 litres/minute which were well above the anticipated and environmentally approved rates. The project was re-designed and re-bid to incorporate a bolted, gasketed primary segmental lining to limit water inflows during excavation and tunnel construction proceeded using two new hybrid TBMs equipped with multiple grouting capabilities. The construction of the tunnels continued to face challenges with grouting due to the variable ground conditions based on a prescriptive approach with grouting as instructed by the designer. This approach proved to be problematic and not sufficiently flexible so further contractual changes were made to the project which relinquished the daily management of grouting operations in the tunnel to the contractor. Completion of these tunnels were completed 9 years after the initial start of construction and highlights that challenges of the ground conditions that were encountered and the overall project restrictions.

The twin 8.5 km Hallandsås Rail Tunnel was sited to transect through a geological horst area with a history of tectonic deformation, high permeability and an elevated groundwater table with environmental restrictions. Initial construction approaches by both drill and blast and TBM methods included "open" excavation with pre-excavation grouting which were unsuccessful to control the drawdown of the groundwater table. Subsequent construction included "closed" mode excavation with a TBM in conjunction with pre-cast lining including a pilot tunnel for ground freezing of a 300 m long zone ahead of the advancing TBM which proved to be the successful solution. Completion of the tunnel was completed 20 years after the initial start of construction and highlights the challenges of such construction risks.

Another interesting project related to the risk of the impact of the groundwater regime is the twin 57 km Gotthard Base Rail Tunnel. Zangerl *et al.* (2008) presents the historical monitoring and evaluation of the consolidation settlements up to 12 cm of the overlying bedrock adjacent to the 57 km corridor of the Gotthard Base Rail Tunnel in Switzerland that was identified as a possible risk for construction of the new project. The magnitude of the settlement in relation to a 2 km deep tunnel excavated in fractured crystalline rock was unexpected and appears to be related to large-scale consolidation resulting from groundwater drainage and pore-pressure changes around the tunnel. Numerical models were able to partially explain the unexpected settlement. The implication of this information is that the stability and safety of overlying infrastructure such as dam/reservoirs needs to be thoroughly evaluated even for deep sited tunnels in bedrock.

12.5 Tunnel construction impacts and disturbances to the community

An often unrecognized risk of tunnel construction is the impact and disturbances that are caused to the local community and the presence of sensitive machinery/banking computer systems during excavation.

Some of the key disturbances include general enhanced noise due to ventilation fans, blasting vibrations, dust generation, and increased truck traffic for muck removal. Mitigation for the impact of noise is commonly addressed with the requirements for silencers for ventilation fans and sound barriers and extended walls around the work site. Further mitigation may include limited or restricted working hours during dayshift only.

Vibrations and noise impacts due to blasting can typically be limited by establishing the site specific blasting characteristics and limiting the size of blasting charges with routine monitoring as confirmation. Blasting times may also be restricted to peak daylight times such as 12.00 noon. The vibrations due to TBM excavation in shallow bedrock have been noted in several project cases to be appreciable and reverberate to overlying residences and offices causing variable disturbances depending on the time of day and commonly result in complaints from the community, particularly during the night.

Careful consideration should be given to potential community impacts during construction and evaluated during the design stage in order to recognize if alternative working sites and tunnel alignments can be modified to limit or prevent any such impacts.

12.6 TBM entrapment and relief

The entrapment of a TBM is possibly the greatest risk for tunnel construction due to the severe impact that can result and associated schedule delay to the project. The typical schedule delay associated with the entrapment of a TBM is 180 days. TBMs may become trapped within very weak rock conditions associated with geological faults due to squeezing of the weak ground around the TBM with direct contact to the TBM cutterhead and/or shield. This causes jamming and the inability to rotate the TBM cutterhead to advance.

The most common mitigation measure for the freeing or release of a trapped TBM is the construction of a bypass tunnel from a location immediately behind the TBM and extending forward by excavation around the TBM, either along the side, or immediately over the TBM, to reach to the cutterhead location of the TBM. Bypass tunnels are used to remove all weak rock conditions from being in direct contact along the TBM cutterhead and/or shield and stabilization of the materials. Stabilization of the ground deformed around the TBM can be performed by groundwater depressurization, grout injection, spiling, as well as by the installation of steel ribs or lattice girders. Bypass tunnels have been successfully completed for numerous tunnel projects constructed using TBMs. Figure 12.3 presents a bypass tunnel constructed along the top of a TBM to allow for the drilling and depressurization of the lahar deposit that was fully saturated and prevented advance of the TBM.

A possible mitigation measure to reduce the risk of TBM entrapment when excavating through a geological fault zone is to intermittently rotate the TBM cutterhead

Figure 12.3 Bypass tunnel over TBM.

during the maintenance shift when extensive rock deformation may occur and come into contact with the TBM cutterhead or shield.

12.7 TBM special problems and design features

While TBMs have successfully constructed numerous tunnels, there has been several serious delays to projects, which were associated with special problems that occurred during construction. One of the most serious problems to experience with the use of a TBM is the failure of the main bearing which is a key component of the TBM for operation. The main bearing of a TBM can fail due to over and extended thrusting due to very high strength rock conditions beyond what was anticipated requiring the TBM to operate at high thrust in order to continue to excavate the very high strength rock. The presence and extent of rock with very high strengths is therefore very important to characterize along the entire tunnel alignment so a proper TBM design with appropriate total power and thrust capabilities can be considered.

TBMs can be designed with special features to address particular risks if identified early during the design. The cutterhead can be sized to be larger than the shield and allow the extension of the gauge cutters to overcut the tunnel to a slightly larger diameter to prevent unexpected deformation of the rock to impact the shield and possibly result in entrapment of the TBM. The typical overcut capability that can be included is about 100 mm on diameter.

TBM cutterheads may be subjected to excessive wear of the steel fabrication where highly abrasive rock conditions are present. Enhanced face plates can be prepared and included in the design to prevent such excessive wear and repair during construction.

As previous mentioned for the use of TBMs in squeezing ground, it is possible to include for increased thruster capacity and increased torque capability of shielded TBMs

12.8 Generation of fine materials during TBM excavation

A key risk during the TBM excavation of both weak and strong rocks is the generation of excess fines that can cause blockage of the cutterhead and conveyor systems requiring regular cleaning and therefore significantly impacting overall production. TBM excavation of weak and altered rock can lead to the generation of a significant amount of clay materials. In addition, the TBM excavation of some strong rocks can lead to the generation of cohesionless fine materials that enter into the main bearing and result in serious damage.

Over-thrusting and over-torqueing during TBM excavation due to inexperienced TBM operators is common and should be closely monitored during the early stages of tunnel excavation. The data logging of TBM performance data has become standard industry practice and the close monitoring and regular evaluation of such information in relation to the generation of fine materials should be undertaken during the early stages of TBM excavation.

12.9 Probe drilling

Probe drilling is a very effective mitigation measure that provides additional information during tunnel construction. Probe drilling is the completion of a percussion (non-core) drill hole ahead of the advancing face of the tunnel to detect the presence of

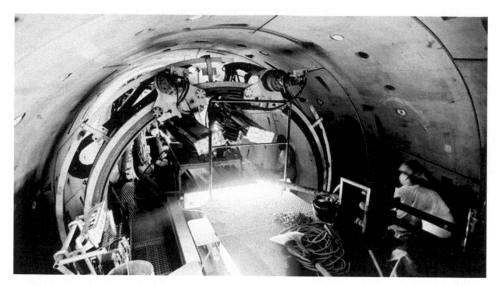

Figure 12.4 Probe drills within shielded TBM.

significant groundwater and/or weak rock conditions. Probe drill holes are commonly completed to 30 m ahead of the tunnel face for drill and blast excavated tunnels and 50 m for TBM excavated tunnels. The penetration rate or rate of advance of the probe drill hole should be accurately monitored and documented during the drilling with the attendance of an engineering geologist or tunnel engineer together with observations of any outflows of groundwater and in particular colour changes of groundwater outflows from clear to brown or grey that may be associated with highly weathered rock conditions. The result of probe drilling should be documented in longitudinal sketches or drawings during construction and discussed with the tunnel constructor to convey the possible risks ahead and develop plans for mitigation measures accordingly. Figure 12.4 presents probe drills included within the shield of a TBM.

12.10 Pre-drainage

Large groundwater infiltration can have a significantly adverse impact on tunnel excavation and stability as well as the safety of workers. Very high groundwater pressures can have a significant impact of the ability for a TBM to advance due to the torque required.

Large groundwater infiltration can prevent the installation of rock bolts in conjunction with grout injection or the placement of grout cartridges. Similarly, large groundwater infiltration will prevent the effective application of shotcrete.

The most technically and cost effective method to manage groundwater infiltration is the use of pre-drainage by the completion of drill holes to intersect rock fractures and fracture zones or geological faults and channel groundwater infiltration through a system of controlled piping away from the tunnel face.

Pre-drainage hole should be completed from a location about 10 m behind the tunnel face and extend to distances up to 50 m ahead of the tunnel face if discrete fracture zones or geological faults are inferred to be present in order to direct the groundwater infiltration away from the tunnel face area in order to allow tunneling activities to continue in a safe manner under conditions with less infiltration.

Upon the successful pre-drainage of groundwater infiltration from ahead of the tunnel face it is necessary to have high capacity construction pump available to discharge the water out of the tunnel face area to the tunnel portal. It is also good practice to excavate a sump or pit within the tunnel floor at regular intervals along the tunnel to allow the channeled groundwater to accumulate in a localized area from which it is then pumped via piping to the tunnel portal.

The pre-drainage of acidic groundwater for any extended duration should utilize stainless steel or fiberglass piping since some corrosion can be extended to occur.

Construction of the twin, 9 m size, 9 km long mine access tunnels at the El Teniente mine in Chile as part of the mine expansion required excavation through an approximately 2 km of highly altered rock zone that contained an elevated groundwater table with a measured pressure of about 11 bars. As excavation advanced towards this zone a series of long horizontal drain holes were drilled to provide pressure relief and to facilitate pre-excavation for ground improvement and the control of water infiltration. Figure 12.5 presents the completion of a 300 m long drain hole and ongoing drainage from the hole.

Figure 12.5 Groundwater drain holes at El Teniente.

12.11 Pre-excavation grouting

Pre-excavation grouting is the injection of cementitious materials into rock fractures performed ahead of the advancing tunnel face. Pre-excavation grouting is commonly performed prior to the intersection of weak rock conditions and/or fracture zones typically containing large amounts of groundwater that have been identified by probe drilling or other investigative techniques. Pre-excavation grouting has a significant impact on tunnel advance rates since the tunnel excavation works of drilling and blasting or TBM excavation must be stopped in order for the grouting works to be performed.

Pre-excavation grouting should be considered to be performed if there exists a risk of the intersection of weak rock conditions commonly associated with major geological faults and significant groundwater infiltration due to the prevailing hydrogeological conditions along the tunnel alignment and in particular where the tunnel alignment is sited below the groundwater table.

Pre-excavation grouting should be compensated for when it is performed on a unit rate basis and include compensation for the set-up or hook-up of the grouting equipment, quantity of grout injected defined by volume or weight, and stand-by time for the tunneling crews, as tunnel excavation is typically interrupted during grouting operations.

An appropriate quantity of grouting materials should be maintained at the project site at all times during construction to prevent any delays from having to perform grouting in order to continue to advance the tunnel.

Pre-excavation may be required to be performed where environmental regulations prohibit any impact to the local groundwater regime such as within nature reserves or where the groundwater resource is used for supply for local drinking or agricultural uses.

Pre-excavation grouting is typically much more effective for reducing groundwater infiltration in comparison to post-excavation grouting since the tunnel is acting as a drain whereby the groundwater infiltration typically changes location.

Figure 12.6 illustrates a typical set up for pre-excavation grouting in a drill and blast tunnel as at the powerhouse access tunnel at the Alto Maipo Hydropower Project in Chile.

12.12 Post-excavation grouting

Post-excavation grouting is the injection of cementitious materials into rock fractures performed behind the advancing tunnel face after the onset in groundwater inflows. Post-excavation grouting can be focused on discrete locations where high volumes of grouting infiltration occurs.

The distinct advantage of post-excavation grouting is that the work can be performed independently of the tunnel excavation operations thereby not impacting tunnel advance rates.

Post-excavation grouting is however not as effective at limiting groundwater inflows as it commonly only results in the re-location of the groundwater inflows and not complete cut off. Post-excavation grouting is most appropriately implemented to reduce major groundwater inflows at discrete locations to manageable volumes to allow tunneling works to proceed.

Figure 12.6 Pre-excavation grouting activities.

12.13 Pilot tunnels

The construction of pilot tunnels has proven to be a very successful mitigation measure when poor rock conditions have been encountered in a medium to large sized tunnel and the support of such conditions presents a challenge. Pilot tunnels are commonly adopted upon the intersection of major geological faults or other significant sections of challenging ground conditions including highly fractured rock with high groundwater pressures.

The construction of a pilot tunnel allows for the investigation of the length and nature of the ground conditions with a small tunnel that can be supported and the stability maintained in a much more controlled manner. A pilot tunnel can also serve to drawdown the groundwater regime and therefore decrease groundwater pressures thereby contributing the overall improved stability of the area. A pilot tunnel can also serve to allow for the pre-treatment and pre-stabilization of the problematic section including grout injection and/or spiling etc.

The construction of a pilot tunnel and evaluation of the rock conditions encountered serve to provide key information for the design of a solution for the enlarged final tunnel through challenging ground conditions. Figure 12.7 presents an example of a pilot tunnel excavated as part of a major highway tunnel project.

12.14 Investigative techniques during construction

The intersection of adverse rock conditions during tunnel construction typically warrants the collection of additional information about the nature and extent of the adverse conditions in order to be able to evaluate the conditions and develop a design solution to allow tunnel construction to proceed with minimum delays.

Figure 12.7 Pilot tunnel.

Standard techniques of long horizontal geotechnical drilling can be utilized from within the tunnel or from specialized drilling niches to collect samples of the adverse conditions to define the three-dimensional geometry of the conditions. The geotechnical drill holes can serve as pre-drainage and also possibly to allow grout injection. For shallow depth tunnels it may also be acceptable to perform additional investigative drilling from the surface into the zone of interest. Geotechnical drillholes completed from surface may be used for pumping wells to attempt to depressurize the zone of interest if high groundwater pressures are present. A thorough evaluation of geotechnical information should be compiled to develop an interpretation of the geometry and anticipated behaviour of the area as part of a design solution.

Innovative technologies including geophysical surveys can also be performed during tunnel construction to investigate the conditions ahead of an advancing tunnel. Tunnel Seismic Prediction (TSP) developed by Amberg Technologies (Dickmann & Krueger, 2014) allows for the forward identification of fractured rock conditions and can be used both for drill and blast and TBM excavated tunnels. The technique is capable to investigate up to 250 m ahead of the tunnel face and requires to stoppage of all work activities in order to limit any noise interference with the data collection and therefore is commonly performed during the maintenance shift during tunnel construction. TSP has become a commonly utilized tool for risk management during tunnel construction to plan mitigation measures and minimize overall project delays. Figure 12.8 illustrates the layout and principle of the tunnel seismic prediction method.

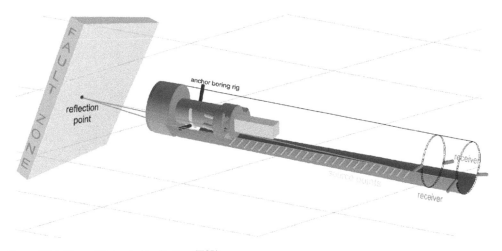

Figure 12.8 Tunnel Seismic Prediction (TSP).

12.15 Additional tunneling equipment and resources during construction

One of most effective mitigation measures for a rock tunnel project is to utilize additional equipment and resources during construction. For a rock tunnel project where appreciable risks have been identified as part of the design, or only encountered during construction, it is prudent to adopt a construction planning approach whereby tunnel excavation is to be carried out from both ends of the alignment and/or from intermediate access locations in order to reduce the risk of delays in the event that one of the construction teams does not achieve the anticipated progress due to unexpected construction risks such as adverse subsurface conditions.

This mitigation strategy should be adopted for all long and deep tunnels as well as where significantly adverse conditions are present along the tunnel alignment. In such cases, it is prudent to identify practical locations for intermediate access adits to facilitate additional construction teams. Intermediate access locations may include both adits and shafts in areas of low topographic cover in order that the construction of the intermediate access can be completed in a timely manner and not represent a critical path construction activity for the project. It is important that any intermediate access locations not be sited within adverse subsurface conditions.

The deployment of additional equipment and resources during construction requires careful planning as additional construction infrastructure and utilities can be expected to be required including power and water. Additional site facilities will also be required to be established including offices, camps, shops, and water treatment systems.

Chapter 13

Construction cost estimation for rock tunnels

13.1 General

The estimated construction cost for a proposed tunnel is one of the most important aspects for a project. It is required by the client in order to establish the amount of funding to be designated for the project and/or confirm economic feasibility for cost sensitive projects. The costs associated with tunnel projects are related to many site specific factors and time related assumptions and are therefore strongly dependent on production rates. Tunnel construction costs are commonly underestimated due to optimistic assumptions and misrepresentations.

The level of effort and presentation of information for tunnel construction costs should be commensurate with the level of study or project definition that has been completed. For example, a detailed cost estimate should not be undertaken and presented as such for a conceptual study, but rather a benchmark estimate should be presented.

Specialized commercially available computer software has been developed and is commonly used by tunnel constructors to prepare construction bids. However, the estimate of tunnel construction costs can also be prepared using standard spreadsheet software.

Tunnel construction cost estimates should only be performed by qualified individuals with extensive experience in cost estimating specifically for underground construction due to the highly specialized nature of the construction work. While there exist several accreditations for costing engineers and estimators, there does not exist any form of accreditation for costing for underground projects.

Clients commonly ask consultants for early construction cost estimates and in some cases to provide an estimate based on a telephone conversation. It is important for clients to recognize and appreciate that any construction cost estimate should only be considered to be representative if an appropriate amount of effort has been expended to produce the estimate. Similarly, tunneling practitioners should not offer to provide any such estimates or "ballpark" numbers as part of a telephone conversation recognizing that the first cost estimate provided may be disclosed in the public domain or used in an economic analysis and relied upon by the client in a meaningful way.

Tunneling practitioners are encouraged to engage tunnel constructors during the early stages of design for projects to provide a reliable estimate of construction costs for tunnel projects as an alternative to a consultant based construction cost estimate. This approach is very applicable for complex tunnel projects and may only be acceptable for private funded projects but not for publicly funded projects due to a perception of an unfair advantage. Tunnel constructors should not be expected to undertake a cot estimate for a project given the amount of effort required and if a meaningful estimate is to be provided.

Since the construction cost estimate is one of the most important aspects for a tunnel project, it is imperative for the client to recognize that an appreciable amount of time and effort should be allocated for the completion of a representative estimate for the given design stage.

Finally, construction cost estimates for tunnel projects are typically developed during the early stages of project development and are primarily based on a design bid build project delivery method. For design build projects it is important to recognize that additional costs can be associated with the total project cost, which depends on the risk allocation as well as the design costs of the design build team. Dutton *et al.* (2011) discusses the challenges of cost estimates for design-build tunnel projects.

13.2 Costing standards and recommended procedures

The completion of construction cost estimates for tunnel projects should ideally follow some form of industry guidelines to be able to reference the general principles adopted so the results provide confidence to clients for important decision making.

The American Association of Costing Engineers (AACE) have developed a cost estimate classification system defining five classes that serves as an excellent set of guidelines to apply the general principles of estimating for the completion of cost estimates for construction projects and these guidelines are applicable for tunnel projects (AACE Recommended Practice 18R-97, 2011). Figure 13.1 presents the AACE Cost Estimate Classification System. (Copyright © 2011 by AACE International; all rights reserved. Reprinted with the permission of AACE International.)

The AACE cost estimate classification system is based on the following important characteristics:

- Level of project definition (expressed as a percentage of total project definition);
- End usage (typical purpose of estimate);
- Methodology (typical estimating method);
- Expected accuracy range (typical variation in low and high ranges), and;
- Preparation effort (typical degree of effort relative to least cost index).

The level of project definition should be consistent with the level of study or project design as well as the purpose of the cost estimate as follows:

- Class 5 Estimate – Concept Design (< 2%);
- Class 4 Estimate – Feasibility Design (1–15%);
- Class 3 Estimate – Budget authorization (10–40%);
- Class 2 Estimate – Preliminary Design (30–50%), and;
- Class 1 Estimate – Final Design (50–100%).

Different methodologies are available for cost estimating including "top-down" benchmark approaches and "bottom-up" detailed approaches. While top-down benchmark approaches are considered to be applicable for the early stages of design, detailed bottom-up approaches should be adopted for the later stages during preliminary and

	Primary Characteristic	Secondary Characteristic		
ESTIMATE CLASS	**MATURITY LEVEL OF PROJECT DEFINITION DELIVERABLES** Expressed as % of complete definition	**END USAGE** Typical purpose of estimate	**METHODOLOGY** Typical estimating method	**EXPECTED ACCURACY RANGE** Typical variation in low and high ranges
Class 5	0% to 2%	Concept screening	Capacity factored, parametric models, judgment, or analogy	L: -20% to -50% H: +30% to +100%
Class 4	1% to 15%	Study or feasibility	Equipment factored or parametric models	L: -15% to -30% H: +20% to +50%
Class 3	10% to 40%	Budget authorization or control	Semi-detailed unit costs with assembly level line items	L: -10% to -20% H: +10% to +30%
Class 2	30% to 75%	Control or bid/tender	Detailed unit cost with forced detailed take-off	L: -5% to -15% H: +5% to +20%
Class 1	65% to 100%	Check estimate or bid/tender	Detailed unit cost with detailed take-off	L: -3% to -10% H: +3% to +15%

Figure 13.1 AACE costing matrix.

final design. Detailed bottom-up approaches should not be adopted during the early stages of design even if commercially available software or spreadsheet templates are available since the information provides a false sense of the level of detail of the completed design.

13.3 Key assumptions for construction cost estimates

A series of key assumptions are required to be documented as part of any tunnel construction cost estimate as follows:

- Project location (urban/remote – transport of workers and staff/mobilization);
- Tunnel size and length;
- Labour rates (union or non-union)
- Anticipated geological conditions and key construction risks;
- Method of tunnel construction;
- Work Schedule (shifts and days per week);
- Electrical power supply;
- Spoil disposal haulage;
- Tunnel production rate;
- Distribution of Tunnel Support and design components;
- Tunnel Final Lining requirements, and;
- Applicable environmental and safety regulations.

The most challenging key assumption for tunnel cost estimates is the tunnel production rate. The selection of a representative tunnel production rate should be based on consideration of previously achieved production rates for similar size tunnels constructed in similar geological conditions using a similar method of construction. As part of the cost estimate, tunneling practitioners should thoroughly review representative production rates from historical tunnel projects to select as part of a cost estimate.

13.4 Direct construction costs

Direct construction costs refer to those costs attributed to specific items to be constructed and production. For tunnels, these typically comprise the following construction components or activities and aspects:

- Mobilization and demobilization;
- Operating of major plant equipment;
- Portal construction;
- Tunnel excavation;
- Tunnel support;
- Grouting, and drainage measures;
- Spoil disposal, and;
- Water treatment.

Direct costs should be based on and consistent with the proposed construction schedule and are commonly presented in terms of labour costs, material costs, equipment costs, and subcontractor costs.

13.5 Indirect construction costs

Indirect construction costs refer to those costs that cannot be attributed to specific items to be constructed or production. For tunnels, they typically comprise the following construction activities and aspects:

- Purchasing or rental of major plant and equipment;
- Maintenance of major plant and equipment;
- General maintenance;
- Field Supervision;
- Quality control;
- General overhead (administration, security;
- External services (survey, consultants);
- Bonds, Insurance, and Taxes, and;
- General expenses (legal, accounting, human resources, IT, permitting, interest and financing, design and corporate).

Project overhead costs are site related costs including costs of site utilities, supervisors, housing and feeding of project staff, parking facilities, offices, workshops, stores, first aid facilities, and plants required to support working crews in different activities. Stolz (2010) provides an extensive discussion on indirect costs and their uncertainty.

13.6 Construction cost contingencies and profits

Construction cost contingencies represent additional costs for unanticipated or risk events that may occur during tunnel construction. The designation of construction cost contingencies can be based on selected percentages of the total of the direct and indirect costs for items including the risks associated for geotechnical, design, bidding, and type of contract. The amount of contingencies is subject to the level of project definition with geotechnical and design risk contingencies decreasing as the design is advanced. For example, an appropriate level of contingency for a conceptual study prior to any geotechnical site investigations may be as much as 50% if there is evidence of adverse subsurface conditions or construction challenges based on completed projects in the area. A series of recommended percentage based contingencies at the end of a final design assuming a comprehensive geotechnical site investigation has been completed and a risk-sharing strategy is incorporated into the contract are presented in Table 13.1.

Table 13.1 Summary of recommended contingencies at final design.

Contingency	Percentage, %
Geotechnical Risk	15
Design Risk	5
Market Competition Risk	5
Contract Risk	5
Total Contingencies	**30**

The amount of the contingencies for geotechnical risk and design risk should decrease as the various stages of engineering design are performed with a greater amount of geotechnical information and design definition. Hence, larger contingencies should be considered for early stages of a project unless similar projects have been successfully completed previously without cost overruns.

Contingencies should be carefully evaluated for each project at each project stage and be based on consideration of all important factors including amount of geotechnical site investigations completed, design effort, current market conditions, probable type of contract to be adopted and risk allocation. Alternatively, construction cost contingencies can be based on the results of a quantitative risk assessment.

The profit margin is subject to the eagerness of the tunnel constructor to win the project, market competition, and the perceived risks of the project by the tunnel constructor. The typical range of profits for tunnel projects varies from 4% during market conditions of limited projects to 12% during market conditions of a large amount of bidding opportunities for tunnel constructors. Given that the international tunneling industry is experiencing an increasing amount of projects, and this trend appears to be ongoing, profit margins can be expected to range from the middle of this range to the upper limit.

13.7 Client's costs

The costs to be recognized by the client for construction of a tunnel project include the following:

- Land Acquisition/Right-of-Way/Property Use;
- Construction Management;
- Environmental Monitoring;
- External Technical Review Board;
- Disputes Resolution/External Legal Advisors, and;
- Financing.

The cost of construction management with a full time technical and inspection team at the project site may range from 8% to 15% of the total expected construction cost since the function and number of team members is dependent on the size of the overall project. A contingency can be applied to the client's costs to also recognize the inherent uncertainty of these costs. The cost of construction management services to be provided by the tunnel design consultant or another designed independent consultant can be expected to increase significantly with any delays to the project schedule.

13.8 Total anticipated tunnel project cost

The total anticipated tunnel project cost is the sum of the direct costs, indirect costs, contingencies, profits, and client's costs. No further contingencies should be applied to the total anticipated tunnel project cost. It is vital that the total anticipated tunnel project cost is completely consistent with a proposed construction schedule.

13.9 Probabilistic analysis of construction costs and geological uncertainty

Due to the high uncertainty regarding tunnel cost estimates and an increasing amount of cost overruns accompanied by growing frustration by clients on many tunnel projects it is no longer accepted by many clients to present a tunnel construction cost estimate as a point estimate value but rather evaluate the anticipated costs with a probabilistic approach. With the availability of computer software such as @RISK (Palisade, 2016) that can incorporate probability distributions to describe reasonable levels of variability of the key parameters and assumptions, it is possible and has now become expected practice by many clients in the tunneling industry to present a construction cost estimate in terms of probability and probable cost.

Reilly and Brown (2004) presents a practical approach referred to as the Cost Estimate Validation Process (CEVP®) for the estimation of tunnel construction costs that requires firstly establishing the base cost without risks, and then the identification and quantification of risks and opportunities to provide a range of probable costs. Tunneling practitioners are encouraged to consider the use of a probabilistic approach for the estimation of tunnel construction costs in order to present the commonly accepted uncertainty of this information. Figure 13.2 illustrates the concept of the cost estimate validation process.

13.10 Integrated cost and schedule risk analysis

Tunnel construction cost is directly related to the construction schedule due to many time related activities for tunneling. In general, a higher construction cost can be expected for a longer construction schedule for a tunnel project. While schedule risk has typically been ignored in assessments of cost risk, cost risk analyses have more recently included attempts to represent uncertainty in time but these analyses usually occurred without reference to the project schedule.

The American Association of Costing Engineers (AACE) have developed a risk analysis to integrate schedule and cost risk for the estimating of an appropriate level

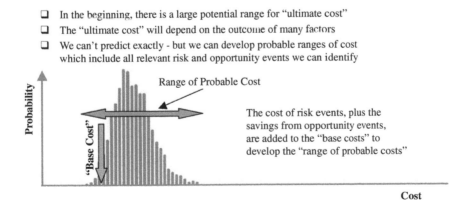

Figure 13.2 Cost estimate validation process.

of cost and schedule contingency for a project in order to include the impact of schedule risk on cost. (AACE Recommended Practice 57R-09, 2011). The approach is based on integrating the cost estimate with the project schedule by resource-loading and costing the schedule's activities. The probability and impact of uncertainties are specified and the uncertainties are linked to the activities and costs that they affect. Monte Carlo techniques are used to simulate time and cost, which allow the calculation of the impact of schedule risk on cost.

This method of risk analysis highlights the identification of cost risks that are risks to the schedule that indirectly extend the use of resources.

13.11 Benchmark comparisons to similar projects

An alternative approach for the detailed "bottom-up" cost estimation of tunnel projects during the early stages of a project is to consider benchmark information from other similar projects. An extensive amount of costing information is available in the public domain on various websites presenting the bid prices for several public projects around the world. A comparison of numerous international tunnel construction costs based on readily available information, including actual bid prices, as well as tunnel construction costs posted on web sites of several international tunneling constructors, has demonstrated similarities of costs between many similar projects in many regions of the world. An analysis of international tunneling costs was completed by Efron and Read (2012) that concluded the costs of tunnel construction in Australia and New Zealand is not statistically more expensive than the rest of the world.

A large database is available based on tunnel bid prices for public projects as it is a common requirement for many government authorities to disclose bid prices. Bid prices serve as a good indicator of the final cost even though they do not necessarily or commonly represent the final construction cost of a tunnel project. Sepehrmanesh et al. (2012) present a planning level cost estimating method using statistical analysis of historical data.

Tunneling practitioners should consider the use of benchmark comparisons for the presentation of tunnel construction costs during the early stages of a project. Clients should also importantly recognize that a benchmark cost estimate also requires a minimum amount of effort that should be expended in order to provide a realistic cost estimate.

Chapter 14

Construction scheduling for rock tunnels

14.1 Identification of key construction activities and graphic presentation

The development and understanding of a construction schedule is an important aspect as part of project and construction management to recognize key construction activities, and the anticipated completion of the project.

The essential method to develop a construction schedule is to construct a model of the project that includes the following:

- A list of all activities required to complete the project;
- The time (duration) that each activity will take to complete based on representative production rates;
- The dependencies between the activities, and;
- Logical end points such as milestones or deliverable items.

Typical work activities to include in a tunnel construction schedule including the following:

- Mobilization;
- Procurement of Major Equipment;
- Preparation of laydowns;
- Set-up of offices and shops;
- Portal construction (or shaft access);
- Intermediate access adits (if included);
- Set-up of equipment and services at portal (ventilation, lighting, water);
- Tunnel excavation and support of all main underground components;
- Ancillary excavations;
- Removal of major equipment (e.g. removal of TBM);
- Removal of services;
- Final lining and tunnel invert;
- Ancillary components (plugs, bulkheads, emergency access);
- Tunnel cleaning, and;
- Demobilization.

Total project schedules commonly include the pre-construction activities such as geotechnical site investigations, design, tendering, and award. Detailed schedules should also be developed for environmental baseline studies, geotechnical site investigations and engineering design in order to track the progress of these important project tasks throughout the duration of the project. Scheduling is commonly a significant role as part of the design of a major tunnel project.

The longest completion path through the completion of all of the work activities is defined as the critical path which is an important aspect to confirm and recognize. While tunnels may form part of a larger infrastructure project where the tunnel is the critical path component of the overall project, stand-alone tunnel projects have their own critical path work activity. This is important to recognize during the early stages of the design in order to evaluate possibilities to reduce the duration of the critical path or alternatives. Figure 14.1 presents an example of a tunnel construction schedule highlighting the critical path.

Float is a common term used for construction scheduling and represents the amount of time that an activity in a project schedule can be delayed without causing a delay to either any subsequent activity or the project completion date.

Commercially available software for project scheduling allow for the development of detailed construction schedules including Primavera P6 (Oracle Corporation, 2016) and Microsoft Project (Microsoft Corporation, 2016). The most frequently used format for the presentation of a construction schedule is a Gantt chart which is a type of bar chart that illustrates the start and finish dates of the main work activities, and shows the dependency relationships between the main work activities.

Rail-line or linear type schedules are also very useful to develop as an alternative method of presentation to a bar chart. Rail-line or linear type schedules provide the benefit of being able to visualize multiple tunnel headings and associated access adits as part of the entire tunnel construction schedule. Figure 14.2 presents an example of a tunnel construction schedule in the format of a linear schedule that includes tunnel construction from multiple adits.

14.2 Procurement lead time for key equipment

The procurement lead time for key equipment for tunnel construction is an important aspect to recognize for the development of a representative construction schedule.

The typical procurement lead time for a new TBM is approximately 12 months based on industry practice from most TBM suppliers. It is uncommon to be able to expect a shorter lead time for fabrication. It should however be noted that there is a large market of used TBMs in the industry which can typically offer a shorter lead time ranging from 6 to 8 months.

In comparison, the typical procurement lead time for a new drilling jumbo is approximately 6 months based on industry practice from most jumbo suppliers. The typical procurement lead time for other commonly used underground equipment including haul trucks, scissor trucks, and shotcrete sprayers ranges from 4 to 6 months. Other important items that may be included for long lead times include rail, piping, and water treatment and concrete batch plants.

The procurement lead time for major underground equipment is influenced by market conditions due to the amount of project activity in the industry. Procurement lead times should be confirmed with suppliers of major equipment for the development of construction schedules during the project design.

Figure 14.1 Critical path construction schedule example.

ID	Task Name	Duration	Start
1	TBM Risk Reduction - Example Project	1157 days	Mon 15-08-31
2	Client Approval of New Approach	10 days	Mon 15-08-31
3	Additional Geotechnical Investigations	145 days	Mon 15-09-14
4	Endose approval	5 days	Mon 15-09-14
5	Planning	20 days	Mon 15-09-21
6	Quotations/Evaluation	15 days	Mon 15-10-19
7	Finalize plans and specifications for drilling	10 days	Mon 15-11-09
8	Finalize contract with Drilling Contractor	20 days	Mon 15-11-23
9	Execute additional drilling	45 days	Mon 15-12-21
10	Evaluate data and report	10 days	Mon 16-02-29
11	Updated TBM risk assessment and project schedule	10 days	Mon 16-03-21
12	TUNNEL EXCAVATION WORKS	1147 days	Mon 15-09-14
13	CONTINUE EXISTING D&B TUNNEL	15 days	Mon 15-09-14
14	Advance existing venture from PK 476 to PK 500 = 24 m	15 days	Mon 15-09-14
15	EXCAVATE TWIN TUNNEL CAVERN	45 days	Mon 15-10-05
16	Excavate twin tunnel cavern	45 days	Mon 15-10-05
17	EXCAVATE MINI-GALLERY TO VENTANA FAULT ZONE	280 days	Mon 15-12-07
18	Excavate mini-gallery to ventbere fault zone - 750 m @ 125 m/month	180 days	Mon 15-12-07
19	Perform preliminary drainage of ventana fault zone	10 days	Mon 16-08-15
20	Excavate mini-gallery through ventana fault zone - 150 m @ 75 m/month	60 days	Mon 16-08-29
21	Perform additional drainage and injection as required - contingency	30 days	Mon 16-11-21
22	TBM ASSEMBLY AT VENTANA PORTAL	415 days	Mon 16-02-08
23	TBM assembly at ventana portal	90 days	Mon 16-02-08
24	Additional modifications to TBM at ventana portal	15 days	Mon 16-06-13
25	TBM delay period at portal	300 days	Mon 16-07-04
26	Move TBM into cavern	5 days	Mon 17-08-28
27	First preparation of TBM to start - walk through fault zone	5 days	Mon 17-09-04
28	D&B FULL SIZE EXCAVATION IN VENTANA AND FAULT ZONE	450 days	Mon 15-12-07
29	D&B Full size excavation from cavern to fault zone - 750 m @ 75 m/month	300 days	Mon 15-12-07
30	D&B full size excavation through fault zone - 200 m @ 50 m/month	120 days	Mon 17-01-30
31	Contingency for problems of D&B through fault zone	30 days	Mon 17-07-17
32	TBM EXCAVATION CONTINUE TO DOWNSTREAM	397 days	Mon 17-09-11
33	TBM excavation through curve section to downstream - 375 m @ 500 m/month	24 days	Mon 17-09-11
34	Prepare intersection area and bypass curve for D&B T-section upstream	14 days	Fri 17-10-13
35	TBM excavation of main tunnel section downstream - 6000 m @ 600 m/month	300 days	Thu 17-11-02
36	TBM assembly to backout	10 days	Thu 18-12-27
37	Backing out of backup and TBM for upstream excavation	21 days	Thu 19-01-10
38	Reposition TBM and back up for upstream excavation	14 days	Fri 19-02-08
39	Walk TBM thru D&B section to upstream end of fault zone main tunnel	14 days	Thu 19-02-28
40	EXCAVATE T-SECTION OF MAIN TUNNEL	200 days	Thu 17-11-02
41	Excavate T-section of main tunnel - 500 m @ 75 m/month via bypass of TBM section 200 days	200 days	Thu 17-11-02
42	MINI-GALLERY EXCAVATION TO MAIN TUNNEL FAULT ZONE	311 days	Mon 17-01-02
43	Mini-gallery excavation to intersection of main tunnel - 300 m @ 150 m/month	60 days	Mon 17-01-02
44	Full size excavation of main tunnel to fault zone - 300 m @ 700 m/month	90 days	Mon 17-03-27
45	Perform preliminary drainage of fault zone	21 days	Mon 17-07-31
46	Excavate main tunnel through fault zone - 255 m @ 50 m/month	140 days	Tue 17-08-29
47	TBM EXCAVATION - UPSTREAM	230 days	Wed 19-03-20
48	TBM excavation upstream section - 3700 m @ 600 m/month	185 days	Wed 19-03-20
49	Disassemble TBM	15 days	Wed 19-12-04
50	Remove TBM and back up (intake or ventbre)	30 days	Wed 19-12-25

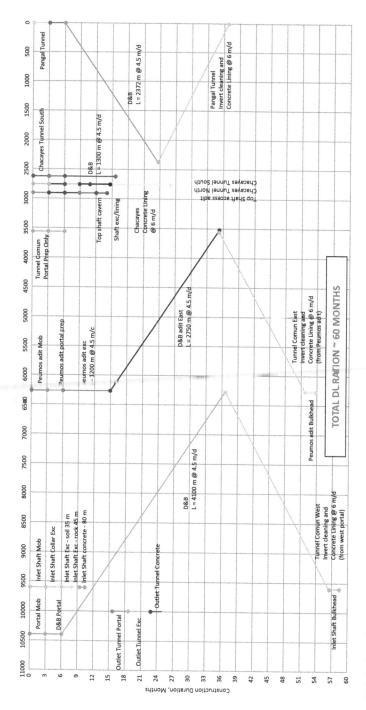

Figure 14.2 Railway line schedule example.

14.3 Evaluation of realistic rates of productivity and working hours

Tunnel excavation and support is typically the critical path activity for most major tunnel projects and therefore realistic production rates should only be considered as part of the development of a representative construction schedule.

Benchmark production rates from historical projects in similar geological conditions based on the total duration of past projects and not based on "best day" or "best week" or "best month" production rates should only be considered as realistic production rates. The use of "best" production rates will result in an optimistic construction schedule.

Tunnel construction is typically performed around the clock and commonly on a 24/7 basis – that is, 24 hours per day and 7 days per week. A tunnel construction schedule should be consistent with the working hours which may be limited by local unions or health and safety authorities.

On some urban tunnel projects, usually the work schedule is limited to 5 days per week in order to reduce the total project costs with no overtime costs. It is therefore important to correctly present the anticipated working hours for the project in the construction schedule.

14.4 Schedule contingencies for risk events

Separate work activities designated as contingencies for risk events including temporary stoppages should be included in a construction schedule based on the evaluation of construction risks. The delay time associated with a tunnel risk event can be highly variable but based on experience in the industry can generally range from 30 days for reduced excavation production through geological faults and other types of difficult ground conditions to 180 days for the entrapment and freeing of a TBM.

Additional contingencies should be considered to be included in a construction schedule for risk events representing delays in procurement of major equipment and for adverse weather conditions which can impact work progress at tunnel portals due to the removal of heavy snowfall or mitigation of avalanches.

14.5 Critical path activities

While tunnels may form part of a larger infrastructure project where the tunnel is the critical path component of the overall project, stand-alone tunnel projects have their own critical path work activity which is important to recognize during the early stages of the design in order to evaluate possibilities to reduce the duration of the critical path or alternatives.

Chapter 15

Tunnel contract strategy and implementation

15.1 General

This Chapter provides an introduction to the main concepts associated with good contracting practices for underground construction including contact types, prequalification, method of payment, risk sharing, geotechnical baseline reports, contact management, partnering, dispute resolution and claims management. Contracting practices are an evolving subject and several improvements and advances have been developed in contracting practices with lessons learned from the completion of past projects from contributions from all stakeholders. Edgerton (2008) provides further details on the foundations for good contracts and best practices for contract provisions. ITA (2011) provides a useful contractual framework checklist for contractual practices for tunnel projects.

15.2 Contract documentation and types of contracts

A consistent and complete set of contract documents is required to be prepared and implemented by all parties with a positive attitude for a successful project outcome.

The series of contact documents that are required to be prepared for pre-qualification and bidding include the following:

- Request for Qualifications;
- Information and Instruction to Bidders;
- Tender Drawings;
- Technical Specifications;
- Form of Contract and or Agreement;
- Dispute Resolution Agreement;
- General Conditions;
- Bill of Quantities and Payment Provisions;
- Geotechnical Data Report (GDR), and;
- Geotechnical Baseline Report (GBR).

A consistency check should be performed both during the preparation of and upon completion of all of the contract documents to identify any contract contradictions.

Several types of contracts are available to consider for the construction of a rock tunnel. A common aspect of tunnel construction contracts is that tunnel constructors are increasingly reluctant to accept any level of geological risk, or at a maximum, only the geological conditions presented and baselined in a GBR. Clients should prepare themselves for such anticipated risk sharing which is considered to represent fair contractual practice in the tunneling industry. In many cases, tunnel constructors will not submit a bid if there is no risk sharing mechanism for geological conditions within the contract.

Typical types of contracts for rock tunnel projects in order of increasing risk to the client include the following:

- Firm Fixed Price;
- Unit Rates and Lump Sums; (with risk sharing);
- Cost-Reimbursable with Incentive Fee (Target Price and Alliance), and;
- Cost-Reimbursable with Fixed Fee.

Figure 15.1 illustrates the relative risk allocation in relation to the typical forms of contract for tunneling projects. The firm fixed price contract places substantial risk on the tunnel constructor but provides cost certainty for the client where the price can be expected to be higher in comparison to other types of contact since the tunnel constructor has to cover for all possible risks. Also, the client only has limited control over the design and construction quality. The tunnel constructor will not wish to accept all the geological risk but rather negotiate a cost-loaded schedule that reflects unit rates for different ground classes; this will allow basis for additional compensation if conditions are more difficult than baselined. A GBR is prepared and included in the contract as a baseline of anticipated conditions. Actual conditions are compared with the GBR as work progresses, and to assess if additional compensation is due. Payments are typically made monthly based on work completed using a negotiated schedule of payments per the percentage of completion of work. The final cost of the tunnel project can be expected to be very close to the initial firm fixed price unless additional compensation is provided. Alternatively, the client may wish that the tunnel constructor assumes all geological risks and a GBR is not required. This approach can be expected to result in a high cost, and higher in comparison to other types of contract.

Figure 15.1 Types of contracts and risk allocation.

A unit rate and lump sum type of contract represents the most common type of contact for tunnels. Unit rate payment items are established for work items with potential for variation in quantity along with lump sum payment items for quantifiable work items. The tunnel constructor is responsible for production and includes profit in the unit rates and lump sums of each item. The final price of the project is dependent on the quantities needed to carry out the work. A GBR is prepared and included in the contract as a baseline of anticipated conditions. Actual conditions are compared with the GBR as work progresses, and to assess if additional compensation is due. Liquidated damages can be assessed as a cost per day penalty for each day that the schedule extends beyond the agreed completion date. An early completion bonus can also be included.

The cost-reimbursable type of contract with incentive fee represents a target cost that is negotiated along with a target fee typically as a percentage of target cost, and includes either a fixed or variable fee (profit) usually as a percentage of the predicted total cost. The target cost and target schedule reflect the baseline of anticipated conditions. If the baseline conditions are better than anticipated, and the tunnel constructor is able to under-run the schedule and cost targets, the client and tunnel constructor "share the gain" of the actual project savings and the target fee is increased. If the baseline conditions are however more adverse than anticipated, and the tunnel constructor over-runs the schedule and cost targets, the client and tunnel constructor "share the pain" of the additional project costs and the target fee is decreased. If the baseline conditions are substantially unchanged and the tunnel constructor over-runs the target cost or schedule, then the tunnel constructor's fee reduces by a pre-negotiated amount. Both a maximum and minimum fee need to be established and agreed on between the client and tunnel constructor. The target schedule is adjusted accordingly on a monthly basis based on the comparison of the actual distribution of rock conditions versus the baseline distribution. The amount of the agreed fee paid can also be linked to key performance indicators of quality and safety. The final cost and schedule of the cost-reimbursable with incentive fee type of contract are uncertain. The baseline conditions have to be agreed on between the client and the tunnel constructor and be representative of the available information so there is no significant bias to favour either party.

The cost-plus fixed fee type of contract is an open book approach where the tunnel constructor is paid for all direct costs (including all site personnel) and an agreed offsite overhead and profit.

The client assumes all the risks of construction and the tunnel constructor is paid at cost to deal with all geological conditions encountered whether within or beyond the baselines with the fee adjusted upward for changes in the encountered conditions and extending the baseline schedule. There is no incentive for early completion and the final cost to the client is uncertain until completion of the project.

The choice of the type of contract is usually made by the client and their internal procurement or purchasing procedures which in some cases are both risk and schedule adverse. This may lead to a preference of a firm fixed cost contract with all risks allocated to the tunnel constructor. However, there is an increasing interest to adopt the most financially attractive type of contract due to limited public funding in many countries and pressure for projects to be economically feasible to private financial targets.

The all-risks contract with a firm, fixed price is considered to be inconsistent in relation to the long term operational risks for hydraulic tunnels and therefore represents poor practice. For hydraulic tunnels this type of contract is at risk of the implementation of inadequate designs for initial support and final lining and the client has only limited control of the final quality of the project to prevent long term risks to operations.

Clients should recognize that the type of contract to be adopted for a planned rock tunnel project can influence the attractiveness of the project for bidders. If a fair type of contract with risk sharing provisions is adopted then tunnel constructors will not shy away from bidding. Contracts that are however perceived to shed or place unnecessary or unfair risk on the tunnel constructor will result in a reduced number of bidders.

The design-build (DB) delivery method can be executed as an engineering, procurement, and construction (EPC) contract whereby the client can retain the responsibility for certain elements of the design that are uncertain for construction such as ground support and hydraulic requirements for hydropower tunnels, and require the tunnel constructor to be responsible for detailed and final design for structural elements. This form of contract is often referred to as a hybrid design-build or hybrid EPC type of contract and allows the client to have control of the final design of the tunnel support during construction which is consistent with good industry practice particularly when payment for tunnel support is re-measured.

A challenge exists for the fair payment of tunnel excavation in rock when a shielded TBM with pre-cast segmental lining is used, whereby direct observations of the encountered conditions are not possible on a regular basis during tunnel construction. Indirect methods may however be applied for the purposes of confirming ground conditions ahead of an advancing tunnel. Bieniawski et al. (2012) present the use of the specific energy of excavation (SEE) as an indirect parameter that can be measured and has been found to be correlated to rock mass quality from three projects.

15.3 Pre-qualification

Pre-qualification should be performed for all tunnel projects to ensure that prospective bidders are suitably qualified to perform the work. Tunneling work is specialized, and the expertise of the management and supervisory field personnel can be critical to the success of the project. If a fair, risk-sharing contract is to be implemented, and is disclosed to the bidders, there will generally be a sufficient number of qualified bidders to provide a competitive environment. In some cases, it may be appropriate for the client to advertise the project to prospective tunnel constructors.

Pre-qualification will serve to address the general experience of the contracting team in similar types of construction, the financial ability to bond the work, and the ability to assign personnel who have sufficient amounts of relevant experience.

The criteria for evaluation of pre-qualification submittals should be specific and quantitative. This approaches minimizes the risk of a bid protest. For highly specialized types of underground work, certain types of experience should be expected to be within the contracting team and in the key supervisory personnel. The types of experience that might be of interest would include soft ground shaft sinking, hard rock tunnel excavation using TBMs, raisebored shafts, and the installation of steel linings for pressurized water conveyance tunnels. The pre-qualification process must take place sometime before the eventual Notice to Proceed and therefore it is typical practice to request a primary candidate for each specific supervisory position as well as one or two alternate candidates.

All pre-qualified bidders should be required to attend a mandatory site visit along with an inspection of representative geotechnical information including drillhole cores.

15.4 Form of payment

Provisions for payment are an intrinsic and important part of a tunnel construction contract and definition of the work and materials for payment should be clearly presented with clear and concise descriptions for each payment item.

For an all-risks contract with a firm fixed price the form of payment is typically based on milestones related to the construction schedule and completion of the work.

For the other forms of contract the payment for the completion of the tunnel works is typically based on the definition of a series of unit rates and lump sums for each main activity of work and/or area of work. Examples of unit rates for payment include, meters of excavation advance, tunnel support components, supply of key materials including steel ribs and grout, and volumes of injected grout and treated construction water. Examples of lump sums for payment include mobilization, site establishment and set-up, water treatment plant and settlement ponds, spoil disposal, and demobilization.

The installation of tunnel support and the respective designs that are to be implemented during construction should be based on unit rate prices for each component or practical measurements of the components that can be easily quantified during construction through inspections such as each rock bolt, each m^2 of mesh, each lattice girder, and each m^2 shotcrete. This contract payment approach is consistent with risk sharing and allows for the tunnel design consultant to make modifications to the design if warranted during construction and for the tunnel constructor to be compensated in a fair manner.

In particular, payment provisions for tunnel support should not be based on a designed or defined "Class or "Type" for example where Class 2 is defined as pattern rock bolts with a total of six rock bolts per advance round. This approach, if implemented formally, represents a rigid and non-flexible method of payment that can result in unfair and/or only partial compensation if a single rock bolt is not installed within the designated design Class. Payment provisions for tunnel support based on an as per Class or Type basis may also result in the complication of new pricing during construction which should be avoided.

The provisions for payment have to be consistent with the overall tunnel construction approach. For example, Payment items for an open gripper TBM excavated tunnel can be based on measured quantities with unit rates for meters of excavation, and installed tunnel support in the form of rock bolts, mesh, shotcrete and steel ribs or lattice girders along with any necessary pre-support measures such as spiling where the required support is evaluated based on the exposed ground conditions. However, for TBM excavated tunnels in rock where a pre-cast concrete segmental lining is included it is not possible to inspect the encountered ground conditions other than probe or core drilling ahead of the advancing face resulting in significant impact regarding progress and the construction schedule. For such tunnel construction approaches it is necessary to establish payment provisions that address the main tunnel excavation. These are typically in the form of a fixed price for the total length of the tunnel alignment, including all related costs of maintenance and cutter changes. They could also take the form of, a unit rate for tunnel excavation and lining, as well as a payment item for additional costs to be applied when the geological conditions cause an impact to TBM progress such as at the intersection of major fault zones. The amount of these additional costs to be established as part of the contract can be designated as a risk allowance to be paid when warranted, which is quantified and based on identified risk events or locations such as the number of geological fault where higher costs are incurred by the tunnel constructor. The use of exploration drilling or other methods, such as seismic prediction techniques should not be used as a basis for payment for a TBM excavated tunnel with pre-cast linings due to the difficulty in the interpretation of any such data and its reliability. Another format for payment may include baselines for TBM operational parameters, but this approach will require a close review of all TBM operational data which is also open for interpretation.

15.5 Risk sharing and compensation for differing site conditions

The internationally established practice for the successful construction of tunnels with minimal disputes and claims has been based on the principle of the sharing of the construction risks between the client and the tunnel constructor. The standard practice is defined whereby the risk of the geological conditions is defined as a baseline condition fully allocated to the client, and the risk of tunnel production is fully allocated to the tunnel constructor. When baseline conditions are exceeded, a declaration of differing site conditions is made, and additional compensation is awarded upon demonstration that the exceedance has impacted the production of the tunnel constructor.

In comparison, risk shedding would typically involve the allocation of all risks to the tunnel constructor, particularly for encountered subsurface conditions. The client typically prefers this approach with the objective of reducing claims on the project. However, the tunnel constructor will include contingencies in their bid when risks are allocated to the tunnel constructor, which the client will have to pay whether the adverse conditions are encountered or not. This approach typically results in a significantly higher construction cost of the tunnel project, but it does not reduce the number of amount of claims.

A risk sharing contract approach offers the most cost attractive approach for the client. This approach avoids hidden costs associated with uncertain work items and will minimize the potential for claims during or at the end of construction. In comparison, an all-risks contract with a firm fixed price, commonly results in a very high cost to the client.

The management of the principle of risk sharing is accomplished with the implementation of a Geotechnical Baseline Report (GBR). When the geotechnical baseline conditions outlined in the GBR have been exceeded, and have been clearly demonstrated by the tunnel constructor to have impacted the works, either by increased costs or schedule, a declaration of a differing site condition is warranted, and additional compensation is provided to the tunnel constructor to a value that is consistent with the degree of the impact to the works.

15.6 Geotechnical baseline reports and implementation

Geotechnical baseline reports have become the internationally established practice for the successful construction of tunnels resulting in fewer and smaller valued contractual disputes. GBRs do not prevent disputes but serve to limit the amount of disputes for a differing site condition that is not anticipated. It is paramount that the GBR be compiled in a concise manner. It should present categorical statements of the baseline conditions and not simply factual or interpretative information, or the range of information without baseline statements. The GBR should ideally only contain clearly stated and presented baselines for geotechnical parameters that can be measured during construction and are not subjective and require interpretation as these will be difficult to confirm and compare to the encountered conditions during construction to decide whether or not they truly constitute a differing site condition to warrant consideration for additional compensation. This allows for the introductions of varying opinions as is recognized for the behaviour of tunnel stability.

While the baseline conditions can be established based on any information, the GBR should aim to present the 'average" or "typical" range of conditions as defined from the site investigation program assuming a comprehensive program has been completed. In the event that limited geotechnical information is available due to limitations of completing a comprehensive site investigation program, it remains imperative that a GBR be compiled based on reasonably expected conditions. In the event that a thorough site investigation is not performed, the GBR should not be biased with only the site investigation data and rather present the most appropriate data that is perceived to be representative for the site. A GBR should seek to set appropriate baselines to limit significant claims during construction due to unexpected conditions being encountered and designated as significant differing site conditions.

The principle of the GBR is that a level and fair field is established for all bidders and the bidders are to assume that the geological risk is allocated to the client and additional payment will be provided if exceedances occur during construction. It is the responsibility of the tunnel constructor to clearly demonstrate the time and cost impact of the differing site conditions on the method and approach of construction that are more adverse than in comparison to the baseline conditions. It is also important to fully appreciate that the baseline conditions presented in a GBR represent a contractual baseline and may not in fact represent geotechnical facts and is dependent upon the level of risk and price certainty that the client will accept.

The implementation of the GBR requires the client and the construction management team to closely monitor and document all of the relevant geotechnical information exposed during construction as well as the detailed performance of the tunnel constructor, and in particular, any impact to the tunnel constructor as observed due to a change in the anticipated conditions.

GBRs are applicable for tunnel projects in rock where the exposed rock conditions can be readily inspected and clearly documented that include the use of drill and blast, roadheader, and open gripper TBM methods of excavation. However, for tunnel projects in rock where a shielded TBM in conjunction with pre-cast concrete segments are used, the application of a standard GBR presents a challenge for confirming the baseline conditions and the determination of a differing site conditions due to limited access of the exposed rock conditions.

Given that the ground conditions encountered during mining with a pressurized face TBM, which are increasing being considered and adopted for hydropower tunnels in challenging geological conditions as a risk mitigation approach, cannot realistically be measured or recorded for comparison to a typical geotechnical baseline, an alternative baseline of conditions need to be established based on TBM operational cycles and parameters, advance rates, and cutterhead tool replacement rates. Deviations recorded during construction from the operational parameters would then serve to be the first indication of ground conditions adversely affecting the TBM. These parameters and conditions of contract can be adopted and negotiated prior to the award of a contract. An example of this approach would include the capping of the amount of hours included for cutterhead interventions along with a very detailed set of parameters to be agreed to for the decision-making process related to need for and duration of any cutterhead intervention. Crew hour rates would also be required to be agreed for hours in excess of this as well as a daily impact cost in the event that the allocation of hours be exceeded. This approach should be adopted for tunnel projects using a shielded TBM incorporating a detailed risk management assessment and process such that all parties understand their potential exposure.

An increasing number of hydraulic tunnels are being constructed using this approach and making the implementation of the GBR even more complex whereby the TBM may be operated in EPB closed mode or open mode. For these projects a new approach is required for risk sharing and should include some form of verification through limited inspection of the exposed rock conditions and monitoring of the TBM operations. In addition, it is believed that an alternative form of GBR may be appropriate involving a baseline of EPB closed mode and open mode operations based on the anticipated distribution of geological conditions.

Improvements for practice and recommendations for the compilation of GBRs for rock tunnels based on shortcomings and in particular, the lack of definition of baseline conditions in the industry, is presented by Heslop and Caruso (2013). Debates continue in the tunneling industry as to how much information should be presented in a GBR and the inherent risk of "over-baselining" which allows for an increasing number and magnitude of disputes, rather than an attempt to reduce disputes. It is the opinion of some tunneling practitioners that the only information that should be contained in a geotechnical baseline should be that which can be verified during construction by simple observations and testing and not subject to any form of interpretation. Thompson (2013) presents alternatives and additional information to a GBR for the allocation of risk and management with examples from completed tunnel projects where a GBR was not implemented.

For design-build tunnel projects a two-step approach should be adopted for the completion of the GBR with a designated GBR-B prepared by the client for Bidding of Proposals, and GBR-C for construction and prepared by each bidder based on their method of construction and evaluated by the client and eventually agreed to with the selected tunnel constructor (Essex, 2008).

15.7 Construction contract and scheduling management

The contract management of the construction of a tunnel is an important stage of the overall project to control the project costs and schedule, and implement decisions, including design changes, based on the encountered conditions.

For a design-bid-build approach, the client should engage a team with good industry experience and qualifications for tunnel design and construction management and ideally should comprise the tunnel design consultant rather than an alternative party that was not part of the original design of the tunnel project. The construction management team should include the following key positions:

- Construction Contract Manager;
- Resident Engineer;
- Assistant Resident Engineer;
- Senior Engineering Geologist;
- Shift Engineering Geologists;
- Tunnel Inspectors;
- Office Engineers;
- Tunnel Design Engineer, and;
- Independent Technical Experts.

The number of resources required for the construction management team is subject to the size of the tunnel project and overall site extent and number of excavations where construction may be carried out concurrently and require to be inspected.

The construction management team should closely monitor and document all of the relevant geotechnical information encountered during construction on a full time basis in conjunction with all work activities as well as the detailed performance of the tunnel constructor, and in particular, any impact to the tunnel constructor as observed due to a change in the anticipated conditions.

Tunnel inspectors should complete shift reports of all work activities including any delays and the reasons thereof such as lack of maintenance of equipment and late delivery of key supplies and materials. Information from the inspection shift reports represents key information for claims defense and should be compiled using a database software program that produces summary charts and tables to be reviewed by the senior staff of the construction management team on a regular basis. Where progress of the tunneling constructor is less than expected it is appropriate for the construction management team to perform a cycle time analysis of the performance of the tunneling constructor to evaluate the detailed information and identify shortcomings in the work activities.

Engineering geologists should perform geological and geotechnical mapping of all exposed rock conditions, collect samples for testing, and compile observations on tunnel stability and or deterioration. The Assistant Resident Engineer should perform detailed visits and inspections of all of the underground works on a routine basis, typically 3–4 times per work week, to evaluate the overall stability of the tunnel, and confirm the installed tunnel support is of acceptable quality, or provide instructions for additional support.

The construction management team should document and summarize all relevant information to be able to make ongoing comparisons to the baseline conditions to be aware of potential disputes during construction. All notices of claim should be thoroughly evaluated by the construction contracts team to assess validity for a possible extension of time of the project schedule and for a possible additional compensation.

Scheduling management should be performed by the construction management team on an ongoing basis during construction by making regular (monthly and possibly weekly) comparisons of the contract baseline schedule to actual progress for all work activities in order to identify possible slippage and loss of schedule float as well as possible near critical path activities for the consideration of schedule recovery solutions including additional resources and equipment. Software programs such as Primavera allow for the detailed tracking and comparison of project schedules to actual progress to try to overcome roadblocks to productivity and the loss of project milestones. Construction management teams are encouraged to have open discussions with the tunneling constructor throughout the construction duration with regards to any recognized losses in productivity and resulting schedule slippages in order to find schedule recovery solutions including possible re-designs.

15.8 Partnering

Partnering is a beneficial process of the overall project management for a project with an objective to create an environment for setting goals, cooperation, and resolution of problems, avoiding or resolving disputes, and improving project outcomes. Partnering can be adopted for both design-bid-build and design-build tunnel projects.

A partnering agreement should be developed between the client, consultants, constructors and key project stakeholders and consist of a mutually developed formal strategy of commitment and communication. The development of a partnering charter is typically completed by the senior level management for the parties, so that involvement and commitment at the highest levels are demonstrated and followed downward to and through all levels of each organization involved in the project.

The client should generally exhibit a strong desire to partner in the solicitation and contract documents. The client's offer to partner should also be discussed at the pre-bid meeting with the potential tunnel constructors. The involvement of the tunnel constructor to accept partnering is critical and underscores a key point to demonstrate a mutual interest and desire to make partnering work for it to be successful for the project.

The costs of the activities associated to prepare and implement a partnering agreement should be shared among all parties. The parties should plan to attend meetings and discuss the process and designate leaders who should be tasked for the planning of a partnering workshop.

An external facilitator should be engaged to lead the workshop, assist the parties in preparing an agreement, develop an issue resolution process, and develop a periodic process of evaluation. Through a periodic evaluation, the parties have the opportunity to review the effectiveness of the process, and to take corrective action. By practicing and maintaining a positive attitude and good communication, ideas and concepts can be expected to be discussed and shared with minimal hostility and separation. The process of partnering commonly starts as a good marriage between all of the parties and the process is tested upon during the early disagreements.

15.9 Dispute resolution

The approach to effective dispute resolution for tunnel projects has evolved to comprise the designation of a Dispute Resolution Board (DRB) with the objective to provide an initial but formal and practical process in order to discuss and resolve disputes between the parties among recognized industry experts in order to avoid legal proceedings.

A DRB is typically created by a contractual agreement between the relevant parties, and typically comprises three members. The members are commonly recognized tunneling industry experts with specific experience in one or more aspects of the proposed construction methods for the project, and should not be an employee or an associate of any of the parties.

The selection process for the members of a DRB commonly comprises a nominated candidate by each party and the first two members then nominate a third member who is subject to the approval of each party, and who generally chairs the board. Another approach for selection is the nomination of five candidates by each party, and then the joint selection of the three board members. The three board members then jointly select the chairman.

The DRB is established at the beginning of the project, and is kept apprised as the work progresses, through distribution of progress reports and through periodic site visits, regardless of any disputes. The DRB therefore has the opportunity to hear from all parties how the project is proceeding, and can view the site conditions prior to any disputed conditions.

If a dispute is presented by a party, the DRB has a working knowledge and background understanding of the project as a basis for assessing the merits and financial impacts. As part of the DRB process, a hearing is held whereby each party is given the opportunity, in an open session, to present their dispute and background supporting information and justification. Following the hearing of both sides of a dispute, the DRB prepares a summary of their findings and recommendations. The recommendations are typically non-binding, and most often limited to merit-type arguments. If a dispute is found to have merit, the DRB typically recommends that the parties resolve the quantum portion themselves. If required and agreed by both parties, the DRB's opinions can also be requested to address the quantum of a dispute. Guidelines for the professional conduct of the DRB should be included in the contract, but the detailed procedures are developed by the board members based on their experience.

The costs associated with implementing a DRB are relatively low and are generally considered to be a cost-effective investment to help motivate the parties for the completion of tunnel construction rather the disputes.

The key difference between the DRB process and other dispute resolution methods such as arbitration is that the DRB is involved at the start of the project, and maintains respect among the contracting parties. The presence of the DRB serves to encourage cooperation between the parties, as well as deterrent, rather than an incentive, to pursue disputes.

The benefits of the practice of the DRB process are apparent from heavy civil construction statistics that have shown that the number of cases of disputes have been limited to less than 1% for escalating further.

15.10 Claims management

The implementation of all of the good contract practices for a tunnel project cannot be expected to prevent the occurrence of claims due to the high uncertainty of subsurface conditions for tunnel projects.

Clients should recognize that a significant effort can be expected to be required to address and manage claims during construction and should therefore engage a group of competent and experienced tunneling professionals to support the construction contract management team. In addition, further supporting resources can generally be required as part of the claims management team due to the commonly vast amount of information to be assembled, processed, analyzed, and summarized.

For tunnel projects where a GBR is part of the contract, there is a fundamental requirement for the contract management team to closely monitor the baseline information during construction and perform an ongoing evaluation of the possibility of baseline exceedances and anticipated claims. Under some circumstances, it may be in the interest of the client to adopt a proactive approach for the assembling and processing of project data with the early engagement of external tunneling professionals to review information to identify the possibility of future claims.

Chapter 16

Risk management

16.1 Risk management and practice

The construction of tunnels is associated with some of the highest risks of any type of construction due to the uncertainty of the subsurface geological and groundwater conditions and the inability to confirm all the conditions along a proposed tunnel prior to the construction of the tunnel. Historical tunnel construction has resulted in some serious accidents with damage to construction equipment, adjacent and overlying infrastructure, and also injuries to workers.

Risk management for tunnel projects has become an increasing established practice using various available tools and should be recognized and accepted by all parties as a beneficial process that promotes the implementation of a culture, processes and structures that are directed towards realizing potential opportunities, identify the hazards and minimize their adverse consequences through effective planning and mitigation measures.

A risk management plan for a new tunnel project should be thoroughly planned and scheduled as part of the project execution and include the following key steps and processes:

- Preparation of Contractual documentation;
- Risk Workshops (qualitative and quantitative);
- Preparation of Risk Register with Preliminary Risk Allocation;
- Regular updating of risk register for each stage of design, and;
- Regular risk meetings during construction with further updating of the risk register.

The contractual documentation and information requirements for good risk management practice includes the completion and inclusion of the following:

- Changed Conditions Clause (Differing Site Conditions);
- Full Disclosure of Available Subsurface Information;
- Ground support design;
- Ground Characterization;
- Risk Register.

The inclusion of a risk register as a contract document is not a well-established practice to date but can be expected to be in the future, which will provide greater recognition and acceptance by all parties. Risk workshops should be organized at each stage of the

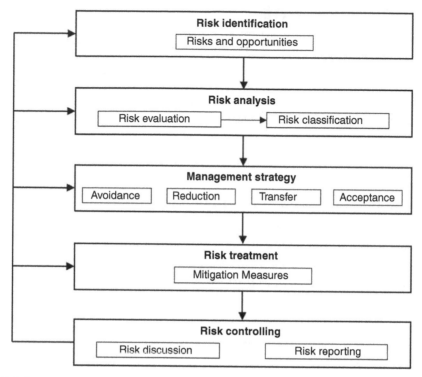

Figure 16.1 Risk management process.

design and be carried out throughout construction as well as operations and should comprise the following key steps as shown in Figure 16.1:

- Risk identification (hazards and opportunities);
- Risk Analysis (probability and consequences);
- Management strategy and allocation (avoid, reduce, transfer, share, accept);
- Treatment of Residual Risks and Verification (mitigation measures);
- Control (discuss and report), and;
- Monitor and Review.

Risk workshops should be facilitated by an experienced risk facilitator and include external technical experts that are well experienced tunneling practitioners who have not been involved with the project design in order to introduce new blood who may be able to identify previously unrecognized risks to the project. Key stakeholders should also be included in each risk workshop.

The presentation and documentation of risk management is commonly documented in a tabular spreadsheet or software format of a risk register. The risk register should clearly include the preliminary allocation of each risk identified to either the client or the tunnel constructor.

Insurance companies have often made decisions to offer or refuse to insure tunnel projects based on limited available information. Insurance companies should engage an independent tunneling practitioner to assist them in performing a comprehensive risk evaluation of a project under consideration for project insurance.

The following codes of practice have been developed as useful references to consider to implement good practice of risk management for a tunnel project:

- ITA Recommendations for Contractual Sharing of Risks (Salter, 1992);
- ITA Guidelines for tunneling risk management (Eskesen *et al.*, 2004)
- The International Tunneling Insurance Group, A Code of Practice for Risk Management of Tunnel Works (ITIG, 2006);
- Geotechnical Risk Management for Tunnel Works, GEO Technical Guidance No. 25, Geotechnical Engineering Office, Hong Kong (GEO, 2009);
- Industry Code of Practice, Underground Mining and Tunneling, MinEx Health and Safety New Zealand (MinEX, 2010);
- Tunnels under Construction: Code of Practice, (WorkCover, 2006) and;
- Guidelines for Improved Risk Management on Tunnel and Underground Construction Projects in the USA, Society of Mining Engineers. (O'Carroll *et al.*, 2014).

16.2 Qualitative risk assessments and risk registers

The initial assessment of tunnel construction risk is usually achieved by means of a qualitative risk assessment and the preparation of a risk register. A qualitative risk assessment is most commonly conducted by means of a risk workshop with the engagement of the senior members of the tunnel design project team along with key members from the client. Key stakeholders of the project may also be included. It is also common to engage external tunneling specialists or technical advisors who have not been involved with the tunnel design. Risk workshops should be managed by a facilitator who has become familiar with the background of the project by means of review of design information completed to date.

A risk register should be designated as a master project document which, upon being created during the early stages of a tunnel project, should be maintained as a live document throughout the life of the tunnel to track risk issues and address problems as they may arise.

There are many different tools that can act as risk registers from comprehensive software suites to simple spreadsheets. A typical risk register should contain the following:

- A risk category to group similar risks;
- A brief description or name of the risk to make the risk easy to discuss;
- The probability or likelihood of its occurrence rated on a number scale (*e.g.* 1–5);
- The impact or consequence should this event actually occur rated on a number scale (*e.g.* 1–5);
- Risk Score (the multiplication of Probability and Impacts), and;
- Ranking (relative position of importance of all risks).

A typical format for a risk register is presented in Figure 16.2 (Goodfellow & Mellors, 2007).

Hazard Number	Hazard	Cause of Hazard	Potential Consequence	Risk Likelihood	Risk Consequence						Risk Score	Control Measures Implemented	Indicators or Metrics	Residual Likelihood after Mitigation	Risk Consequence after Controls						Action Item for Mitigation	Action Completion Date	Risk Owner
					Financial	Schedule	Reputation	Regulatory/Legal	Health&Safety	Environment					Financial	Schedule	Reputation	Regulatory/Legal	Health&Safety	Environment			
1																							
2																							
3																							
4																							

Figure 16.2 Risk register example.

				CONSEQUENCE				
				1	2	3	4	5
				MINOR	MODERATE	SIGNIFICANT	MAJOR	CATASTROPHIC
	SCHEDULE IMPACT TO PROJECT			< 1 week	< 1 month	1 - 3 months	3 - 6 months	> 6 months
	COST IMPACT TO PROJECT			< 0.1 M	0.1 - 1.0 M	1 - 10 M	10 - 25 M	> 25 M
PROBABILITY	1	< 5%	RARE	1	2	3	4	5
	2	5 - 25%	UNLIKELY	2	4	6	8	10
	3	25 - 50%	POSSIBLE	3	6	9	12	15
	4	50 - 75%	LIKELY	4	8	12	16	20
	5	> 75%	ALMOST CERTAIN	5	10	15	20	25

Figure 16.3 Risk matrix for quantitative assessment.

The impacts or consequences should be presented in relation to the commonly important subjects including:

• Financial;
• Schedule;
• Corporate Reputation;
• Regulatory/Legal;
• Health and Safety, and;
• Environment.

The risk score is commonly quantified and presented with a colour coded table with elevated risks shown in orange and red, and low risks shown in yellow and light green. The risk response or action item for mitigation should also include the risk action plan and date to be completed.

Figure 16.3 presents a typical risk matrix with suggested probability and consequence classes with respective descriptions for application to determine a risk score for each hazard as part of a qualitative and quantitative risk assessment. The final results of a risk assessment can be presented in the format of 2D histogram.

16.3 Risk allocation

Tunnel construction risks should ideally be allocated to the party who is both responsible for and can best manage the identified risks. While most clients who build tunnel projects are risk adverse it is important to recognize that a tunnel constructor only has control over his own production of work and the ground conditions in which to build the proposed tunnel. Accordingly, it is common industry practice that the risk of the subsurface conditions is allocated entirely to the client and the risk of construction production is entirely allocated to the tunnel constructor.

As previously discussed, the allocation of the risk of the subsurface conditions is commonly baselined or referenced to a series of assumed conditions and presented in a Geotechnical Baseline Report (GBR) that serves as a contract document accepted by all parties.

16.4 Quantitative risk assessments

Quantitative risk assessments are a simple extension of a qualitative risk assessment and include the addition of quantified consequences in terms of construction costs to allow for a total estimate of the cost impact of construction risks to the project.

The results of a quantitative risk assessment are commonly incorporated into the risk contingency component of the construction cost estimate.

The designated costing values of the consequences in a risk register should be based on past experience of the project team or from consideration of representative projects where similar risks were realized during construction and the actual cost impact was known. Great care must be applied in the development of construction cost contingency values as the total cost impact can significantly accumulate based on this approach.

Quantitative risk assessments are commonly presented in terms of variable values or probability distributions for the construction cost consequences and likelihoods in order to present the overall results as a probability or cumulative distribution to be able to recognize and present the overall probability that a particular total risk costs may be exceeded by at a designated level of probability. Alternatively, the results are presents in terms of a confidence level. For example, a 95% confidence level may be associated with a total risk cost of $500,000 which means that there exists only a 5% chance that the additional costs to the project will be greater than $500,000. This form of risk based information is very valuable for clients to consider and incorporate into their project management budgets assuming the input for the risk register and construction costs consequences are realistic and based on practical experience from the tunneling industry.

Chapter 17

Inspection of rock tunnels

17.1 General

Careful planning is required for a useful and comprehensive inspection of a rock tunnel. The planning for an inspection is subject to the type and operations of the tunnel but is typically of limited extent since it is important to place the tunnel back into full and uninterrupted operation as soon as possible. In order to perform an inspection it may be necessary to place the tunnel completely out of service or perform the inspection during a regularly scheduled maintenance period. In some cases it may be possible to limit the interference of the inspection of operations such as a dual tube traffic tunnel where one tube can be maintained for operations. Tunnels represent key infrastructure that should be maintained in acceptable operating conditions to prevent significant maintenance and repairs. Regular inspections should therefore be performed as part of normal operations.

Tunnel inspections may be performed manually by visual observations made by walking through the tunnel or with the use of remotely operated vehicles (ROVs). Non-destructive testing and inspection technologies are commonly utilized for the inspection of lined tunnels including geo-radar and infrared thermal imaging.

Inspections are commonly performed after many years of operation or when a concern has been observed or identified or believed to have manifested and is required to be addressed for future safe operations so additional or significant damage to the tunnel is not caused from further operations. Whenever an inspection is performed it is important to prioritize the information to be collected but also maximize the amount of information that can be collected as it may not be possible to perform another inspection for an extended period due to the importance of maintaining operations.

Tunnel inspections may include the application of geophysical or other types of non-destructive testing surveys to investigate the condition of tunnel linings. A detailed risk assessment should be performed as part of the planning for a tunnel inspection in order to fully recognize the possible hazards that may exist and what safety actions should be implemented in the event of an incident during the inspection.

Noted references for the inspection, assessment and maintenance of tunnels include the United States Federal Highway Administration (2005) and the Institute of Civil Engineers (2014). Richards (1998) presents international lessons learned and practice from the inspection, maintenance and repair of tunnels including reasons for well planned maintenance and causes for repairs.

17.2 Manual inspections, data documentation, and safety practices

The historical performance and inspection findings of the tunnel should be carefully reviewed in advance to recognize if previous safety related concerns exist for the tunnel. The stability and safety condition of very old tunnels may be marginal due to deterioration of or the absence of tunnel support. Many historical tunnels that were built using masonry linings are associated with the formation of voids behind the lining due to the complete deterioration of the original wooden support.

A large amount of information should be documented during a manual inspection given the extensive planning and associated costs for the inspection. Hence, an inspection team should comprise a minimum number of three to four persons in order that the information can be documented in a timely manner. Observations should be documented not only of any concerns but also of the condition of existing tunnel support. Information to be collected during a manual inspection related to only the civil works should include the following:

- Observation number and time;
- Station location (chainage from portal or established reference);
- Photographs of all relevant observations at each station;
- Observations on tunnel stability and the severity of any concerns;
- Status and condition of tunnel support;
- Observations of deterioration and status of rock and rock type;
- Observations of deterioration of tunnel support – corroded rock bolts, broken shotcrete;
- Observations of condition of tunnel lining (shotcrete or concrete);
- Hammer sounding against tunnel lining to detect possible voids;
- Estimated opening of any cracks and extent of any spalling of tunnel linings;
- Observations on ancillary components of tunnel;
- Observations of any debris along tunnel floor;
- Estimated volume of debris;
- Estimated groundwater infiltration at discrete locations;
- Observations on status and function of drainage system or clogging debris;
- Observations of foreign materials or debris in tunnel (garbage, stones, wood);
- Observations of scour or erosion in hydraulic tunnels;
- Estimated volume of debris in rock trap for hydraulic tunnels;

Geophysical techniques including ground penetrating radar (GPR) may be utilized as part of a manual inspection for the detection and mapping of voids behind a masonry or historical concrete lining.

The lead person involved with a manual tunnel inspection should be a well experienced tunnel engineer capable of identifying unstable or dangerous conditions. The inspection team should be prepared to terminate the inspection in the event that dangerous conditions are encountered, which would prohibit the safe passage of the inspection team.

The air quality in the tunnel should be evaluated in advance by testing at each portal and also monitored during the entire duration of the inspection using an approved instrument. If any diesel powered equipment is planned to be used then additional ventilation can be expected to be required. Existing tunnels with a single point of entry are commonly classified as confined spaces and all local safety regulations should be thoroughly reviewed and adopted as part of the inspection.

The tunnel inspection team may consider to include a qualified miner to perform limited scaling in the event that there exist some locations where potential unstable rock wedges are present, which can be scaled in a safe manner to allow the inspection to proceed. Tunnel rescue teams along with associated emergency safety equipment should be present for the entire duration of a manual inspection of an existing tunnel in the event of an incident. Audio communications using hand-held radios should be used at all times with frequent communication between the inspection team and the portal rescue team at pre-set intervals of time or distance from the portal to confirm the safety and well-being of all members of the inspection team.

All members of the tunnel inspection should utilize the series of standard personal protection equipment for underground construction. All members of the inspection team should be equipped with back up lights as well as light snacks and water in case the duration of the inspection is extended beyond the originally planned duration. Rothfuss *et al.* (2011) presents the requirements for a well-executed inspection of water tunnels. Montero *et al.* (2015) discusses the key aspects of tunnel inspections including the use of robotic systems for inspections.

17.3 Unwatered inspections of hydraulic tunnels using ROVs

Hydraulic tunnels for water supply, and in particular hydropower tunnels, are commonly designated as essential infrastructure that cannot simply be taken out of service for inspections. Hydraulic tunnels are also typically very sensitive to the removal of water or unwatering that may impact its structural condition and therefore such tunnels should not be unwatered unless absolutely necessary. Fortunately, the unwatered inspection of hydraulic tunnels can be easily performed owing to the available technology of remote operated vehicles (ROVs).

The purpose of unwatered inspections of hydraulic tunnels is to confirm the structural condition and overall acceptability for continued safe operations. Unwatered inspections are able to identify exposed rock areas, the possible presence of defects including voids or spalled shotcrete or concrete, the presence of rock or other types of debris accumulating along the tunnel floor, the condition of rock support and final linings that were installed as part of the original construction, and to determine whether additional support is warranted in case of pre-mature or unexpected scour and/or erosion.

With hydropower or other types of water conveyance tunnels, unwatered inspections are typically performed for newly constructed tunnels shortly after a limited period of operations such as within the warranty period of the project to confirm the adequacy of their performance and to allow early repairs to be made before the start of extended operations. However, there exist numerous aged hydraulic tunnels that have never, or rarely, been inspected, particularly after decades of operation. Unwatered ROV inspections should be performed for such aged hydraulic tunnels as part of routine maintenance programs during operations and ROV inspections can be completed within relatively short time periods to limit total outages. The unwatered inspection using ROVs for a hydropower or other types of water conveyance tunnels should be planned well in advance of a planned outage so all of the necessary arrangements can be prepared carefully and in relation to the available outage of operation. Overall, inspections of hydraulic tunnels should ideally be performed at a maximum of 5 year intervals regardless of normal operations for the early detection of any possible damage such that maintenance and repair can be planned to limit possible exacerbation of such early damage and significant defects. Operating authorities of hydraulic tunnels should recognize that the insurance policies for their assets may require regular inspection and maintenance to be performed in order for the policies to remain valid.

The unwatered inspection of hydropower and other water conveyance tunnels has advanced significantly over the past decade and can be expected to continue further with technological advances with remote operated vehicles (ROVs) and ancillary data collection tools including high resolution cameras and videos, and high resolution sonar imaging. ROVs may be operated with a tether or untethered and incorporate a host of data acquisition tools.

Figure 17.1 illustrates a high resolution photograph of a fallen rock block inside a hydropower tunnel.

Unwatered inspections are ideally performed during no-flow conditions during a limited time period for a maintenance outage of operation. It is however also possible to perform an unwatered inspection during flowing conditions but this approach is associated with greater risks due to possible ROV power consumption when travelling against the flow.

The main engineering tasks that should be performed as part of a ROV inspection include the following:

Figure 17.1 High resolution photograph from unwatered tunnel inspection.

- Review all relevant as-built information and compile summary of longitudinal profile of all risk locations and develop preliminary list of key locations for inspection;
- Review geological information for the tunnel including rock types, presence of alteration, nature, orientation, and width of geological faults and shears, and orientation of main rock fractures per rock type;
- Review tunnel hydraulics – distribution of flow velocities for all size cross sections;
- Review problems experienced during original construction – fault zones etc.;
- Review installed support, especially at main geological faults;
- Review previous inspection reports – observed scour/debris/lining conditions etc.
- Review historical hydraulic operations and any headlosses that have occurred;
- Review status of rock trap and history of any cleaning;
- Review information on any debris that may have been removed from tunnel;
- Review type of data and imagery to be recorded and provided by ROV contractor
- Pre-ROV inspection meeting with ROV contractor and AES
- Attendance during ROV inspection with close observations of "live" inspection video and sonar information to identify and instruct for additional logging of information either by video/camera
- Evaluation of ROV data with development of 3D imagery files for updated condition assessment as follows, and;
  - Plan, profile and cross section images;
  - Profile of as-inspected lining type distribution for comparison to as-built;
  - Images of concrete and shotcrete lining locations and transitions;

Figure 17.2 High resolution sonar image from unwatered tunnel inspection with ROV.

- Images of any types of defects observed such as cracking, spalling of lining;
- Images of volume and locations of debris/fall-out;
- Images of geometry and locations of scour/erosion;

- Compilation of tunnel structural condition report - DRAFT AND FINAL

The as-built construction records for the tunnel where an unwatered inspection is planned should be thoroughly reviewed to confirm the probable construction geometry and the originally installed tunnel support and lining. This is done to identify possible locations of interest where instability may be present. For an unwatered inspection to be performed during flowing conditions it is important to review the hydraulic conditions of the tunnel and the expected distribution of flow velocities in relation to the as-built tunnel cross sections along the tunnel alignment.

Figure 17.2 illustrates an example of a high resolution sonar image produced an ROV inspection and illustrating the integrity of a concrete lined section of a tunnel.

Chapter 18

Renovation, repairs, and decommissioning

18.1 Renovation of rock tunnels

Existing rock tunnels may be required to be renovated, which is sometimes referred to as rehabilitated, for an alternative use from their original design function or to be enlarged for alternative future operations. Typical examples include the enlargement of existing heavy freight rail tunnels for double-stacking of rail cars and historical rail tunnels for bicycle, walking/hiking routes or for new infrastructure utilities such as gas, water or sewage pipelines or fibre optic communications. Renovations are generally referred to as geometrical modifications to a tunnel that are made that are not required for structural reasons for future operations. However, in many cases it is necessary to also modify or enhance the structural integrity of a tunnel that is subject to renovations since a change in geometry can often require increased tunnel support and lining measures.

The renovation of an existing tunnel is typically performed by taking the tunnel out of service in order to allow the new works to be completed in a timely manner without undue interruptions. However, in some cases it may be necessary, and is possible, to perform the renovations during pre-defined, and often, very limited time windows, during ongoing operations.

The extent of renovations that have been performed on existing tunnels varies widely and includes the following:

- Minor enlargement of tunnel profile (roof, sidewalls, or invert) for increased clearance in unlined rock tunnels;
- Lowering or modification of tunnel invert for longitudinal drains to convey groundwater infiltration out of the tunnel and prevent ponding or icing;
- Installation of a tunnel floor for improved access and passage for alternative functions such as a bicycle tunnel, and;
- Construction of emergency access or egress passages including an entire parallel small sized tunnel with connections to the main tunnel.

The renovation of existing, and in particular historical tunnels, should be based on a comprehensive review and full understanding of the original design and construction. A thorough inspection should be performed of a tunnel planned for renovation including specific destructive and non-destructive testing of the materials (shotcrete or concrete) of any existing linings. Figure 18.1 illustrates the enlargement of the Weehawken Rail Tunnel, which was renovated based on the removal of the original masonry lining

Figure 18.1 Renovation of Weehawken Rail Tunnel.

and construction of an alternative final shotcrete lining design that incorporated a waterproof membrane.

18.2 Repair of rock tunnels

Repairs are commonly performed in rock tunnels that have been out of services for an extended period such as a historical access adit or have been operating for several years and where no or limited maintenance has been completed. Repairs are generally referred to as those works related to the structural improvement or enhancement of a tunnel in order to improve the long term safety and stability for future operations.

As with tunnel renovations, the repair of an existing tunnel is typically performed by taking the tunnel out of service in order to allow the new works to be completed in a timely manner without undue interruptions. However, tunnel repairs commonly represent a smaller amount of total work and have often been completed during pre-defined time windows during ongoing operations.

The extent of repairs that have been performed on existing tunnels varies widely and includes the following:

- Installation of rock bolts and shotcrete or other forms of support to improve the stability of the tunnel;
- Installation of a waterproofing system in conjunction with a tunnel liner to prevent groundwater infiltration or leakage into the tunnel;
- Installation of a tunnel lining along discrete sections by shotcrete or concrete;
- Installation of insulation systems to prevent build-up of ice in cold regions;
- Drilling of drainage holes to prevent water pressure build-up behind an existing shotcrete or concrete linings;
- Grouting of voids and deteriorated timber blocking behind existing concrete linings;
- Complete replacement of existing masonry lining with shotcrete or concrete, and;
- Major structural enhancements as part of seismic upgrading.

The repair of existing, and in particular historical tunnels, should be based on a comprehensive review and full understanding of the original design and construction. A thorough inspection should be performed of a tunnel planned for repairs to confirm the prevailing conditions and stability status. Consideration should be given to performing destructive and non-destructive testing of the types of support including drillcores from the existing linings (shotcrete or concrete) and pull-tests on existing rock bolts. Figure 18.2 shows extensive repairs that were completed including the

Figure 18.2 Repair of hydropower tunnel.

installation of steel ribs following multiple collapses at the Rio Esti Hydropower Tunnel.

18.3 Decommissioning of rock tunnels

Although most tunnels constructed within the past century continue to be used in their original function, or have been renovated or repaired, there may be a requirement to temporarily or permanently decommission a tunnel when renovation or repair is deemed too costly or practical to meet current safety regulations for continued operations.

The requirements for decommissioning of existing tunnels may be subject to government regulations due to environmental considerations and the prevention of any future impact or contamination to the environment. However, in general, there does not exist any set procedures and requirements for the decommissioning of tunnels and each case should be thoroughly evaluated taking into consideration the purpose of the decommissioning and any possible future impact that decommissioning may cause. The entire historical function and operations of a tunnel that is planned for decommissioning should be evaluated to understand the past loading conditions that have been subjected to the tunnel.

Historical tunnels that are planned for decommissioning and have been out of service for an extended period should be inspected by either remote methods if the safe and stability is a possible concern or by direct observations in order to understand the current conditions and integrity of the tunnel. Historical tunnels that have been out of service for an extended period may be associated with numerous locations of instability and deterioration and therefore may be at a high risk of future collapse. For such prevailing conditions it may be necessary to firstly re-stabilize certain sections of a tunnel to prevent any future instabilities from occurring particularly if new infrastructure may be planned adjacent to or overlying the existing tunnel.

Historical tunnels that are flooded may be drained in a controlled manner if possible by respecting local environmental requirements to allow for a manual inspection, or inspected using a submersible type of remote operated vehicle (ROV). However, it is important to recognize that the draining of a flooded tunnel may result in further instability.

The decommissioning of an existing tunnel commonly requires the effective sealing or plugging of the ends of the tunnel to prevent future access. The effective sealing or plugging at the ends of a tunnel will depend on the existing conditions and geometry of the tunnel portals and entrance sections including the overlying portal slopes as well as the long term loading conditions that may be imposed by sealing or plugging. If portal canopy structures are present it will be necessary to integrate the existing structure with the new seal or plug in order not to demolish the existing structure that may cause instability. If a simple rock portal is present it will be necessary to integrate the new seal or plug into the surrounding rock. The traditional design criteria for seals or plugs are based on structural loading and prevention of shear failure, and hydraulic leakage. A key consideration for the design of a seal or plug for a tunnel is the potential for long term corrosion due to acidic groundwater conditions. Morald and Kolenda (2008) present the main considerations for the design and construction of tunnel and shaft plugs for both temporary and permanent requirements.

In some cases it may be worthwhile to include future access through a seal or plug in order to be able to perform inspections to confirm the long term condition of the tunnel after decommissioning and to be able to re-enter for routine inspections, maintenance or repairs.

Chapter 19

Case histories and lessons learned

19.1 General

Tunneling is an experienced based profession and tunneling practitioners can learn a significant amount of information from previous projects, particularly those tunnel projects associated with challenges and problems where solutions were implemented for successful completion. Lessons learned from several historical tunnel projects are presented in the following sections. The discussion of these past projects is not intended to disclose confidential information or embarrass the tunneling practitioners who may have been involved on these projects but rather to highlight some important lessons learned for future tunnel projects.

19.2 Lesotho Highlands Water Project Phase 1, Lesotho

The Lesotho Highlands Water Project Phase 1 comprised a 45 km water transfer tunnel through sub-horizontally layered basalt rock that was originally designed as an unlined tunnel. Extensive pre-construction mineralogical and swelling testing was performed on drillcore disc cut samples known to contain expansive constituents including zeolites and smectites. Based on the results of drillcore testing the risk of large scale deterioration was recognized. Extensive deterioration of the select layers of basalt was observed during the early stages of tunnel construction and was designated as "crazing" comprising the development of micro-fractures. Upon recognition of the large extent of deterioration of the basalt layers during construction, a decision was made to adopt a concrete lining over the entire length of the 45 km transfer tunnel. Figure 19.1 shows the deterioration of basalt that was identified during construction.

Lesson Learned: The deterioration of moisture sensitive rocks can be expected to be scale dependent with greater deterioration occurring on larger samples or on a large scale around the entire profile of a tunnel as observed during construction in Lesotho. Multiple layers of basalt were exposed along the 5 m size TBM excavated tunnel which resulted in relaxation of many areas containing expansive minerals such as laumontite, which resulted in deterioration. Where the long term stability and deterioration of suspect rock types are uncertain for the acceptability of an unlined tunnel design, full-scale tests or trial excavations should be constructed during the design stage to fully evaluate and recognize the risk.

Figure 19.1 Deterioration of highly amygdaloidal basalts.

19.3 Pacific Place Pedestrian Tunnel, Hong Kong

The development of the Pacific Place Shopping Center in the early 1990s included a pedestrian tunnel connecting to the nearby Admiralty subway station. The pedestrian tunnel was aligned under Queens Road which is a major thoroughfare in Hong Kong that includes the tram line. The original tunnel support design for the pedestrian tunnel was based on rock bolts and shotcrete. However, during construction the depth of weathering to competent rock varied but was typically deeper than expected and rock bolts installed along the tunnel roof penetrated into the overlying highly weathered rock. The deep weathering of the rock required a change of the tunnel support design to incorporate steel ribs over the upper half of the tunnel profile and founded halfway along the sidewalls in good quality rock. The new tunnel support design was required to be conservative and include steel ribs over the entire length of the tunnel in order to prevent any possibly collapse during construction. Figure 19.2 illustrates the installation of steel ribs for final support of the pedestrian tunnel.

Lesson Learned: The depth of weathering of rock in tropical and sub-tropical environments can be expected to vary significantly and extend to great depths. The design of shallow tunnels in urban areas where the risk of a collapse could be extremely serious with the loss of life should include a comprehensive site investigation to profile the entire tunnel alignment and depth of weathering. Tunnel support designs for shallow tunnels in urban areas where deep weathering is present and uncertainty exists should be based on a conservative approach to limit excavation advance and require high capacity passive support.

Figure 19.2 Steel sets for shallow pedestrian tunnel support.

19.4 Taipei Ring Road Tunnels, Taiwan

The Northern Second Expressway was constructed incorporating twin 3-lane road tunnels through the low lying hills along the eastern side of Taipei during the early 1990s. The Chungho Road tunnel was excavated through historical coal mine workings including cohesionless fire clay adopting a top heading and bottom bench approach. Significant deformation and yielding of steel rib supports occurred during the excavation of the bottom bench in the first tube. Less tunnel support was installed in the second tube and a major collapse occurred as well as additional damage to the first tube upon excavation of the bottom bench of this tunnel (Brox & Lee, 1995). Figure 19.3 illustrates some of the temporary major support installed shortly after the collapse.

Lesson Learned: The excavation of large span twin tube tunnels sited in weak rock should be staggered to prevent any influence between the tunnels. The excavation stability and performance of the installed tunnel support of the early advancing tunnel should be thoroughly evaluated during construction and modifications should be made to the tunnel support design as warranted prior to the excavation of the subsequent tunnel tube.

19.5 Bolu Mountain Road Tunnel, Turkey

Extreme deformation and damage of the initial tunnel support occurred during construction of the twin, 3.2 km, 16 m wide, Bolu road tunnels of the Anatolian Motorway

Figure 19.3 Emergency support works for large tunnel collapse.

Project. The tunnels were sited among thick bands of highly plastic clay fault gouge sandwiched between a metamorphosed limestone and marble formation and excavated with a top heading and bench. During construction the allowable deformation under the new Austrian Tunneling Method (NATM) design approach was increased from 200 mm to 350 mm and finally to 500 mm. The ongoing acceptability of large deformation, which at many locations, exceeded the design criteria up to 700 mm, resulted in significant damage to the initial support comprising steel ribs in conjunction with 25 cm of 20 MPa shotcrete (Brox & Hagedorn, 1998). Additional deformation typically occurred upon excavation of the bottom bench. The excavation sequencing was subsequently changed to include a curved invert in conjunction with the top heading bench as well as sidewall drifts to limit large deformations which proved to be successful. Figure 19.4 shows the significant deformation and resulting damage of the initial tunnel support that occurred during construction.

Lesson Learned: The design and construction of large tunnels in very weak rock conditions should be based on an appropriately conservative approach for excavation sequencing with an ongoing performance based evaluation of tunnel stability based on instrumentation data reviewed on a regular basis by well qualified and experienced tunnel engineers.

19.6 Gotthard Base Rail Tunnel, Switzerland

The preliminary design stage of the twin, 57 km, Gotthard Base Rail Tunnel included the construction of a 5 km, 5 m diameter TBM excavated exploration tunnel to

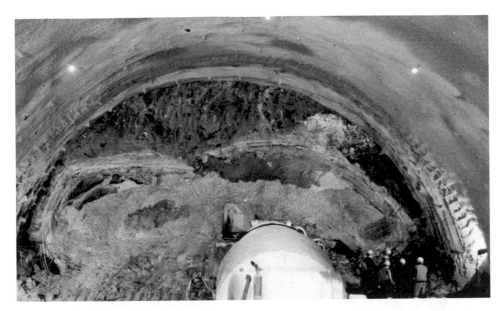

Figure 19.4 Severe deformation and damage of large tunnel.

investigate the extent and nature of weak rock conditions along the alignment of the main tunnels. The exploration tunnel served to allow geophysical surveys and drilling investigations to be completed and pre-treatment grouting injection to be performed as mitigation measures prior to the construction of the main tunnels. Figure 19.5 illustrates the extreme overstressing that occurred under a rock cover of 1700 m during the excavation of the exploration tunnel.

Lesson Learned: Exploration or pilot tunnels should be incorporated into the early design stage of a tunnel project where key geological risks have been identified and may be difficult to investigate using traditional methods including deep drillholes. Exploration or pilot tunnels are an effective means to support the de-risking of a tunnel project.

19.7 Seymour Capilano Twin Drinking Water Tunnels, Canada

The Seymour Capilano drinking water tunnels comprised twin, 7.2 km long, 3.8 m diameter, TBM excavated tunnels with a maximum cover along the tunnel alignment of 600 m. Extensive laboratory rock testing was performed indicating a wide range of rock strength varying from 35 MPa to 260 MPa with an appreciable amount of tests below 100 MPa. Overstressing of the tunnel profile increased as TBM excavation advanced under deeper cover including the occurrence of rockbursts with the ejection of the rock blocks under the deepest cover. The occurrence and degree of overstressing was not foreseen as part of the design of the project nor presented as a possible construction risk in the geotechnical baseline report as no in situ stress measurements were performed as part of the pre-construction site investigations. The observations of the overstressing have been validated using a prediction technique for the overstressing

Figure 19.5 Extreme overstressing of rock at 1700 m depth.

of deep tunnels (Brox, 2012). Figure 19.6 illustrates the significant overstressing that occurred under a rock cover of 550 m in altered rock during the excavation of the tunnels.

Lesson Learned: Pre-construction site investigations of medium to deep tunnels in rock should include in situ stress testing using hydraulic fracturing via drillholes or overcoring in test excavations to provide key information on the state of in situ stress in order to evaluate the risk of overstressing, design applicable tunnel support systems, and convey all related construction risks to bidders.

19.8 Niagara Hydropower Tunnel, Canada

The excavation of the Niagara Hydropower Tunnel was completed using a 14.5 m diameter open gripper TBM. The tunnel alignment was sited in a high horizontal in situ stress environment and aligned under a maximum cover of 150 m in weak rock with an average rock strength of 25 MPa. Significant overstressing and fall out of rock extending to a depth of 5 m occurred immediately behind the fingershield of the TBM during excavation that required the installation of spiling to maintain tunnel stability that resulted in significant delays to the project schedule. Figure 19.7 illustrates the significant overstressing and resulting fall out that occurred under a rock cover of 150 m in the low strength rock during the excavation of the tunnel.

Figure 19.6 Severe overstressing of rock at 550 m depth.

Figure 19.7 Significant overstressing of rock at 150 m depth.

Lesson Learned: A comprehensive evaluation of the potential of overstressing for tunnels sited in low strength rock and within a high in situ stress environment using representative site information should be performed as part of early design and include a careful review of observations and instrumentation data from full scale test excavations.

19.9 Arrowhead Inland Feeder Water Transfer Tunnels, USA

The Arrowhead Inland Feeder Water Transfer Tunnels comprised the 3.7 m diameter, 13 km Riverside Badlands Tunnel, and the 9.3 km Arrowhead East, and the 6.4 km Arrowhead West Tunnels, both of 5.8 m diameter. All three tunnels were excavated through highly variable and high pressure water bearing metamorphic and granitic as well as weakly cemented sedimentary rock conditions with fines using shield TBMs operating in open mode with no face pressure in conjunction with the placement of non-bolted and non-gasketed pre-cast concrete segments as initial support. Although the highly variable and challenging ground conditions were recognized as part of the design, more difficult conditions were encountered during construction requiring pre-excavation grouting as directed by the designer using both chemical products and micro-fine cements, which caused significant delays to TBM production. The Riverside Badlands Tunnel was completed with challenges ahead of schedule and under budget. However, the construction of the Arrowhead tunnels resulted in the drawdown of the groundwater table which was prohibited within the forestry reserve and required further pre-excavation grouting with increased impact to TBM production. The segmental lining design was also changed to a bolted and gasketed approach to prevent subsequent seepage along the tunnel and allow for second stage contact grouting. In addition, further changes were introduced for the operations of the TBMs to enhance performance including mixing arms in the plenum and a conditioning system to deliver foam and other conditioning agents to the plenum and the screw conveyor. Figure 19.8 illustrates the pre-excavation grout ports incorporated into the TBM shield.

Lesson Learned: The impact to the groundwater regime should be recognized as a key risk for rock tunnels sited in mixed and weak rock conditions at depth under high water pressures due to the draining effect of excavation. The mitigation approach for such challenging conditions to achieve realistic and acceptable construction schedule is to adopt the design principles used for soft ground tunneling with active and controlled face pressure as offered with earth pressure balanced type TBMs.

19.10 Canada Line Transit Tunnels, Canada

The Canada Line Transit Tunnels comprised twin, 6 m diameter, 2.2 km subway tunnels excavated using an earth pressure balanced (EPB) TBM through massive sandstone and mixed overburden conditions including fill under the downtown core of the transit alignment. Cut and cover tunnels using both a stacked and lateral arrangement were constructed along a separate 2 km commercial district along the overall transit alignment. A comprehensive geotechnical site investigation program was completed prior to construction based on a high frequency of drillholes commensurate for a major urban tunneling project. Construction of the tunnels through the

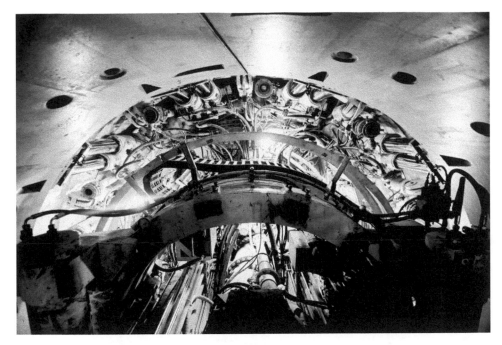

Figure 19.8 Pre-excavation grouting with TBM excavation.

downtown section was successfully completed without any major problems and delays. Construction of the cut and cover section required an extended duration due to the temporary diversion of utilities and the scope of required work, which resulted in interruption and loss of business to many owners along the commercial district. The decision to adopt cut and cover tunnels versus bored tunnels along the commercial district was based on the perceived risk of the subsurface conditions and potential impact to the project schedule. The subsurface conditions along the commercial district were of better quality than the conditions along the downtown section. Figure 19.9 illustrates the completed pre-cast lined tunnel.

Lesson Learned: A detailed evaluation of tunnel construction risks and constructability in relation to cost and schedule impacts should be performed as part of early studies using local tunneling experts familiar with the site specific subsurface conditions in order to confirm the feasible construction methods that can be adopted to minimize and prevent interruption to the community and business owners.

19.11 Ashlu Hydropower Tunnel, Canada

The Ashlu Creek Hydropower Project comprised a run-of-river intake structure, a 135 m, 3 m diameter raisebore intake shaft, a 4.4 km, 4 m diameter open gripper TBM excavated tunnel, and a surface powerhouse. The project site was located within a large U-shaped glacial valley with steep walls of exposed granitic rock. The tunnel was aligned sub-parallel to the valley with a maximum cover of 550 m. The only form of pre-construction geotechnical investigations that were completed included field

Figure 19.9 Pre-cast concrete lining – Canada Line.

geological mapping. A total of six geological faults were predicted to be encountered, most oriented sub-perpendicular to the tunnel alignment. A sub-parallel oriented geological fault was intersected along 48 m of the tunnel requiring the installation of steel ribs over the entire length. Overstressing occurred upon TBM excavation under increasing cover of about 250 m at the 10 o'clock to11 o'clock and 4 o'clock to5 o'clock positions viewed downstream coincident with the high stress locations in relation to the sidewall of the valley. An increased amount of rock support was required to contain the fractured rock as a result of overstressing. Brox *et al.* (2008) presents details of the construction of the hydropower tunnel. The tunnel was constructed under a design-build fixed price all risks contract. Figure 19.10 shows the 48 m long section of steel rib supports for the stability of a major geological fault zone.

Lesson Learned: Pre-construction site investigations should have been completed to provide information for design and evaluation of construction risks including the prediction of overstressing such that appropriate provisions could have been included with the TBM for the safety of workers.

19.12 Forrest Kerr Hydropower Project, Canada

The Forrest Kerr Hydropower Project comprises a run-of-river intake structure, a 10 m size, 4.5 km tunnel, and an underground powerhouse site in fair to good quality

Figure 19.10 TBM tunnel support for long geological fault.

volcanic rock. Two access tunnels were constructed to provide permanent access into the underground powerhouse. All excavation was undertaken using drill and blast methods. The original design of the portals for the access tunnels to the underground powerhouse located them within topographic gullies that were natural drainage channels and where deep overburden was present. Immediately prior to the start of the excavation of the portals, modifications of the locations of the portals were made to shift the portals to adjacent ridges that comprised competent rock conditions for improved stability and minimum excavation requirements. Figure 19.11 shows the modified location of the powerhouse access portal.

Lesson Learned: The design of tunnel portal locations should be thoroughly evaluated as part of the design in order to identify all possible risks that could impact construction costs and schedule. In general, tunnel portals should not be sited within topographic depressions that are commonly associated with deeper weathering and poor quality rock conditions but rather within rock ridges that are commonly associated with good quality rock conditions.

19.13 Rio Esti Hydropower Tunnel, Panama

The Rio Esti Hydropower Project comprises a regulating dam and diversion structure, a 6.5 km diversion canal, reservoir intake, a 10 m size, 4.8 km, pressure tunnel and a

Figure 19.11 Tunnel portal location along small ridge versus gulley location.

surface powerhouse. The operating head of the hydropower plant was 112 m with a flow of 118 m³/s with an installed capacity of 120 MW. The project is located among horizontally layered young and low strength volcanic sedimentary rock, particularly including poor quality red tuffaceous beds, with a depressed groundwater table. The final tunnel support and lining design comprised various thicknesses of shotcrete as the planned project schedule, and the original project economics, could not accommodate a final concrete lining. After nine years of operation multiple collapses occurred along most of the tunnel alignment comprising the fall out of large slabs of shotcrete lining. An initial headloss was detected however the hydropower plant was continued to be operated for another year before increased headlosses and eventual shutdown to investigate the collapses. Frostberg *et al.* (2007) presents details of the completion of construction. Figure 19.12 illustrates the deterioration and fall out of weak volcanic rock that was subjected to first time saturation during hydropower operations.

Lesson Learned: The first time saturation of weak young volcanic rock and fluctuating internal pressures during hydropower tunnel operations represents a key risk for long term operations for a shotcrete lined tunnel whereby additional loading conditions are developed, and if not recognized as part of the tunnel support design, may, and in this case, result in multiple collapses along the tunnel. The design review process of the design-build project should examine all possible construction and operation risks. The contract duration was too short to design and construct in accordance with the prevailing high risk geological conditions to allow for the placement of a concrete lining for long term operations. In addition, upon any indication of a headloss, the tunnel operations should be stopped immediately and an unwatered inspection performed using an ROV.

Figure 19.12 Tunnel sidewall collapse of non-durable volcanic rock.

19.14 Chacayes Hydropower Tunnel, Chile

The Chacayes Hydropower Project included a 2 km transfer tunnel to convey water from a secondary run of river intake to the main river intake. The secondary run of river intake was located within the nature reserve where an access was not allowed to be constructed. The construction of the transfer tunnel was undertaken using an open gripper TBM to provide the main access to the secondary intake location. The open gripper TBM daylighted out of a sub-vertical rock cliff located near the intake. Figure 19.13 shows the large alluvial boulders encountered in the tunnel during TBM construction.

Lesson Learned: The 2 km transfer tunnel was aligned through a small low-lying ridge separating the two intakes that was assumed to comprise rock but where no pre-construction site investigations were completed. A river channel deposit comprising boulders was encountered during TBM excavation through the rock ridge resulting in delays due to tunnel instability and special support measures to be installed. Unanticipated conditions can be encountered during construction of tunnels aligned along the side of major post-glacial valleys where historical fluvial deposits can be present.

19.15 Los Arandanos Hydropower Tunnels, Chile

The Los Arandanos Hydropower Project comprises 14 km of tunnels in association with two run-of-river intakes, penstock siphon, and surface powerhouse. The project is located at low elevation in the Cachapoal Valley south of Santiago among a folded syncline of young volcanic rocks. A comprehensive geotechnical site investigations program including a total of 33 drillholes was planned and executed including seismic surveys, deep rotary and sonic drilling, and in situ testing and laboratory testing. Prior to the site investigation

Figure 19.13 Open gripper TBM excavated tunnel with boulders.

program, multiple campaigns of field geological mapping were performed by different consultants. Both campaigns of the field geological mapping did not thoroughly inspect and map the project area due to limited road access within the project site and missed the identification and inspection of key rock units suspected to be non-durable. Figure 19.14 shows red tuffs of low durability identified during field mapping.

Lesson Learned: The completion of comprehensive geological mapping across the entire site in order to identify all possible geological faults and suspect rock conditions including non-durable rock units prior to the planning of site investigations. Adequate budgets and schedules should be allocated by clients for field mapping including the use of helicopter for difficult access areas. Google earth should also be utilized for the identification of key rock units and the planning of access for field geological mapping stations. The discovery of key geological risks after the commencement of a major site investigation program requires re-prioritization of resources and budget.

19.16 Red Lake Gold Mine High Speed Tram Tunnel, Canada

The Red Lake Gold Mine constructed a 6 km long tunnel at a depth of 1500 m as part of an expansion of the mine to a new deep deposit. No geological or geotechnical information was available from the deep area of the mine prior to tunnel construction. Tunnel construction of the 26 m^2 tunnel was by drill and blast methods using standard rebar 2.5 m long rock bolts and mesh. The tunnel encountered very weak rock conditions over a length of about 700 m characterized with a rock quality of GSI = 35.

Figure 19.14 Non-durable red tuffs in outcrop.

Squeezing of each of the tunnel sidewalls of 0.5 m occurred requiring re-excavating of the sidewalls for multiple times, closure of niches and the risk of rupture to the utility services for tunneling. The severe squeezing conditions was monitored using instrumentation and continued for 18 months and caused the breakage of rock bolt face plates and significant damage and cracking of shotcrete with continued squeezing. Excavation of the tunnel through the very weak rock conditions was completed without any serious incidents, injuries or damage to equipment. Figure 19.15 illustrates large deformation and damage to shotcrete support of weak rock subjected to high stresses at a depth of 1500 m.

Lesson Learned: The 6 km long high speed tram tunnel represents a new life of mine conveyance that warranted deep investigations to provide geotechnical information for the design of initial tunnel support prior to construction, to recognize the risks associated with tunneling through very weak rock conditions, for design of long term tunnel support, and for realistic planning of the project schedule.

19.17 Pascua Lama Mine Conveyor Tunnel, Chile

The Pascua Lama Mine is one of the largest gold mines in the world located at an elevation of 5000 m along the border between Chile and Argentina among large glacial ice caps. The development of the new mine includes a 4 km, 5 m wide tunnel for the

Figure 19.15 Significant squeezing of weak rock at 1500 m depth.

conveyance of crushed ore from the open pit to the plant site. The operating conveyor was designed to be suspended from the tunnel roof by way of a rock bolt and chain hanger system. The tunnel alignment was designed in relation to an underground crusher station adjacent to the open pit and traversed below a mountain ridge separating the open pit and the plant site. The tunnel was aligned under shallow rock cover along 1.5 km of the eastern part of the tunnel alignment. Several surface creeks were present immediately upslope of the eastern part of the tunnel with semi-continuous surface runoff as well as subterranean flows. The rock conditions along the tunnel alignment comprised fractured granite with partial mineralization. Construction of the tunnel along the eastern section resulted in significant groundwater inflows that prevented the effective use of grouted rock bolts and shotcrete that necessitated a tunnel support design change to split sets and mesh. Severe corrosion of the tunnel support including softening and deterioration of previous shotcrete was observed during the subsequent construction of the tunnel due to the groundwater acidity of a pH of1.9. Prior to the closure of the project a section of fiberglass rock bolts was installed for future performance evaluation. Figure 19.16 shows high volume infiltration of acidic groundwater during tunnel construction.

   Lesson Learned: The tunnel alignment was designed with a long low cover section immediately below a major groundwater recharge area in a mineralized area. Significant acidic groundwater inflows should have been identified as a major risk to

Figure 19.16 Highly acidic groundwater inflows.

both tunnel construction and long term operations of the tunnel during the design stage of the project justifying a modification to the tunnel alignment along an adjacent mountain ridge and/or pre-drainage mitigation measures as part of the tunnel design.

19.18 Los Condores Hydropower Tunnel, Chile

The Los Condores Hydropower Project includes a 4.5 m diameter, 12 km tunnel excavated using a double shield TBM with pre-cast concrete segmental lining due to non-durable rock conditions. The project layout included an intermediate access adit where the TBM was launched to excavate downstream and upstream sections of the main tunnel. Limited pre-construction geotechnical investigations identified a major geological fault that intersected both the construction adit and the upstream section of the main tunnel. The inferred intersection length along the construction adit was 150 m. The inferred intersection length along the upstream section of the main tunnel was similar based on an optimistic interpretation of limited drillhole information. The key risk for the project related to the launching of the TBM to excavate through the major geological fault along the construction adit and the main tunnel with the potential for squeezing conditions based on limited drillhole information. Figure 19.17 illustrates the double shield TBM used for tunnel construction.

Figure 19.17 Double shield TBM for practical construction and lining of non-durable rock.

The inferred geological interpretations of the major geological fault of significant widths, and the identified key risk of squeezing of the double shield TBM if used through the fault, warranted additional geotechnical investigations during the design stage of the project in order to establish a reasonable construction schedule for the project. The tunnel constructor relied upon the limited geological information and interpretations at bid to base a schedule for early launching but then re-evaluated the risk of TBM squeezing after award that created a great concern for a schedule delay for the project.

Lesson Learned: The significance of the major geological fault intersecting two locations of the project and the risk of TBM squeezing was not fully recognized during design to realize to perform additional investigations to have a better understanding of the construction risks prior to bidding.

Chapter 20

Engagement and roles and responsibilities of professionals

20.1 Engagement of professionals

The engagement of professionals to perform studies and design services should be based on a consulting engineering services agreement. The scope of the services to be performed should be clearly defined and based on a series of tasks with estimated effort of resources and deliverables. Services that are excluded should also be clearly defined along with information to be provided by the client and the overall responsibilities of the client. The expected duration of the services should be noted in the services agreement along with any required milestones.

The engagement of professionals by a client should be based on well proven qualifications in the industry with particular experience for the type of tunnel under consideration. The qualifications and experience of key individuals of the design team should be confirmed with historical references.

The services performed by consulting engineers including engineering designs, construction cost estimates and schedules will be fully relied upon by the client and are expected to be completed to a standard of care in the industry. The services to be performed should be insured against any professional errors and omissions in accordance with good and standard industry practice. It is important to recognize that professional engineering services are not, and cannot be, guaranteed, but are rather insured. The design and construction of underground projects is associated with high risks and any design related errors or omissions commonly results in major costs for the re-work or re-construction. Therefore, the level of the professional liability insurance that should be required to be maintained by the design professionals should be appreciable. Typical insured limits for design professionals are in the order of $2 Million to $5 Million Dollars. Greater insured limits can however be considered subject to the level of risk and precedent of the design for the project and the perceived costs of repairs in the event that damage occurs during the early stages of operations.

Insurance policies that are provided by insurance companies for design professionals are "claims-based" policies which limit any claim against a design professional to the period of the validity of the insurance policy. This is typically until the end of the construction warranty period, which limits any claims that can be made against the design professional in the distant future. It is important that the duration of professional liability insurance should extend to the end of the construction warranty period as well as to a post-completion period until such time that the operation of the tunnel has been confirmed by the client.

Several other relevant terms and conditions should be included in a professional engineering services agreement including the limitation of liability of the design professional. The design professional should carefully prepare a services agreement with support by legal counsel or legal counsel should thoroughly review any proposed services agreement provided by a client. A common misunderstanding by clients in the industry is that professional services by consulting engineers should not be treated the same as the procurement of materials and equipment administered by the purchasing department of a client.

20.2 Roles of professionals

20.2.1 General

Various roles and responsibilities are required for successful planning, design, and construction of rock tunnels. Based on industry practice around the world, there has been some confusion regarding the individual roles and responsibilities of civil engineers with typical limited education and experience in geology, and geologists, typically not involved with design and construction supervision, and their overstepping of their respective roles. It is important that all tunneling practitioners practice within their area of qualification and professional liability. Clients should closely review the experience and qualifications of all project team members for a tunnel project to confirm that appropriate resources, including specialist experience, will form the tunnel team and be available to advance the design in a timely manner to the required project schedule and milestones.

20.2.2 Geologists

Geologists play a critical role for rock tunnels for the planning of site investigations, characterization of all geological, hydrogeological, and geotechnical data for design, and mapping and data documentation during construction. Geotechnical data and interpretative reports should be prepared by geologists as part of the design. These reports should include the compilation of plan drawings and profiles illustrating the distribution of all factual data in conjunction with similar presentations of interpretative drawings and reports. Senior geologists should be designated the responsibility of quality control and assurance of all site investigation data observations and documentation as well as field testing and presentation of results.

Geologists should also be responsible for the compilation of the Geotechnical Data Report (GDR) as part of the contract documentation. The contents of the GDR should importantly include all relevant geological information and a comprehensive review of the contents of the GDR should be performed to confirm that no important information is missed.

During tunnel construction, it is common practice for geologists to be responsible for the mapping of the encountered conditions and all observations for each working shift and for the reporting of all information in a project database system. Geologists are also responsible for the compilation and summary and presentations of all of the encountered conditions typically displayed by way of longitudinal charts and profiles maintained both digitally and on hand produced drawings in the site offices. Geologists are also responsible for interpretations to be made of exploration data collected during construction such as from probe and rotary drillholes completed ahead of the advancing tunnel headings. Compilation of the collected data together with an ongoing understanding and appreciation of the encountered and predicted conditions is vital as part of a good risk management approach in order to identify possible mitigation measures and prevent delays.

20.2.3 Geotechnical engineers

The role of geotechnical engineers for tunnel projects usually encompasses the preparation of the design criteria and basis, the selection of design parameters and material properties, and the analysis of excavation stability and the development of the initial support designs for the portals and tunnel based on information documented through various memorandums and reports. Geotechnical engineers are commonly responsible, and typically the primary author, for the compilation of the Geotechnical Baseline Report (GBR) in conjunction with many other members of the project team.

Geotechnical engineers are also typically responsible for the analysis of the stability of the portals and the support requirements as part of the overall portal design, which often includes the presence of overburden materials.

20.2.4 Civil engineers

Civil engineers are typically the work horses of the tunnel design and construction team. They are usually responsible for the implementation of the applicable national or international standards of design practice for the development of design drawings incorporating the required drawing standards in conjunction with AutoCAD designers. Civil engineers will also contribute to the compilation of the design criteria and basis. Civil engineers are commonly responsible for the design of drainage requirements at portals, access roads, spoil sites, hydraulic requirements including pipelines and pumps, concrete structures, and temporary works as required. Civil engineers are commonly also responsible for project scheduling and cost estimating.

20.2.5 Tunnel engineers

Tunnel engineers are commonly referred to as those practitioners who have several years of experience and are designated senior members of a tunnel design and construction team. Senior tunneling practitioners are typically civil, geotechnical, or mining engineers who have been involved with many major tunneling projects. They typically serve as design managers, deputy project managers or project managers who are responsible for the management and coordination of the design and review of all of the contract deliverables. Tunneling engineers will typically be the main players involved with risk assessments and the development of risk registers.

20.2.6 Independent technical experts

Independent technical experts are commonly engaged by the client from the outset of the planning stage. They are involved during the entire course of the project through all stages of design and during construction. They provide independent opinions on all technical issues including proposed site investigations, design criteria, project delivery and contract approach, design, scheduling and costs, and construction management including claims and risk avoidance. A typical group of at least three independent technical experts form and represent a Technical Advisory Board (TRB) who report directly to the client but should be allowed to communicate and respond directly with the design team.

Independent technical experts usually comprise individuals who have gained extensive experience in the underground industry and may include ex-consultants and ex-contractors. Many independent technical experts have specialist experience, for example, for TBMs or unlined or traffic tunnels, and such specialist experienced members should be engaged by the client according to the nature of the project works.

20.3 Responsibilities and liability of professionals

The responsibility and liability of professionals involved with a tunnel project is important to recognize and appreciate for the interest of all parties. The design engineer, which is commonly an established consulting engineering firm, is the single entity responsible for the design of the project, and importantly, the implementation of the design during construction, and is designated as the Engineer of Record. Accordingly, the design engineer should perform regular site inspections during construction to confirm that the design is being constructed and implemented correctly. For tunnel projects it is common practice that the design engineer is given the opportunity by the client to have a design representative at the project site on a full time basis as part of the construction management team.

It is important for the client to recognize that the design engineer, as the designated Engineer of Record, has full authority over the design and design implementation. Under a design-bid-build project delivery approach, the client has the opportunity to review and influence the design to confirm that all the client's requirements are included in the design as well as changes made during construction. In comparison, under a design-build project delivery approach, the client does not have the opportunity or authority, to influence, change, or prevent changes to, the design, and must accept the design as prepared and implemented by the tunnel constructor's designer.

At the end of tunnel construction, the Engineer of Record will be expected to provide a design conformance statement, which normally, will comprise the confirmation that the construction of the tunnel has been completed in general conformance with the issued design. This design conformance is typically recognized as the "sign-off" by the Engineer of Record. The design conformance statement is a document prepared by the Engineer of Record and issued as part of the record or as-built documentation for the project and generally assumes that all defects or non-conformances identified have been repaired and addressed during construction. The Engineer of Record is responsible that the design, and its conforming construction, meets the functional requirements and overall intended purpose for the tunnel. The design professional can, in some cases, be responsible for latent defects during the defect liability period, even if regular site inspections were performed during construction.

Notwithstanding an implied design life for the project, the duration of the responsibility and liability of the Engineer of Record continues only to the extent of the professional liability insurance period, which is commonly limited to the period of the services provided, or to the end of construction. However, as noted earlier, it is in the interest of the client to require that the professional liability insurance of the design professional be extended to post-completion to such time when the early operations of the tunnel have been confirmed. The period of the limitation of liability of a design professional may however be subject to the legal jurisdiction of the project and in some cases may extend to a long period after the completion of construction. The extent of the period of the liability of a design professional can however be limited as negotiated within a services agreement. The design professional should carefully evaluate the professional liability insurance requirements for a given project and seek legal advice and review before proceeding with any design services.

Chapter 21

Health and safety

The health and safety practice for tunneling constructors is commonly governed by local authorities and regulations while the health and safety practice for tunneling practitioners is typically governed by the safety procedures established at the site by the tunneling constructor.

Guidelines for good occupational health and safety practice in tunnel construction has been published by the International Tunnel Association (ITA) Working Group 5 (ITA, 2008) that provides a framework of regulations and guidance to ensure that underground construction is performed in a safe manner. Some detailed guidelines for the safe working in tunneling for tunnel workers and first line supervisors has also been published by the International Tunnel Association Working Group 5 (ITA, 2011). Additional useful references on the subjects of refuge chambers, ventilation, and compressed air work are available from the ITA website.

Tunneling practitioners will commonly be required to visit the project site during the early stages of design including during the execution of a site investigation. During these early stages of the project before construction begins, tunneling practitioners should be equipped with the basic personal protective equipment (PPE) including safety vest and comfortable boots. Safety boots may be in appropriate and actually a hindrance to wear in mountainous areas during the inspection of tunnel portals and other areas as well as for field mapping. The safety policy of the client for PPE should be reviewed and confirmed prior to all site visits.

During construction tunneling practitioners should be fully equipped with all necessary personal protective equipment (PPE) for working underground including safety boots, hard helmet, high visibility vest or jacket, cap lamp, self-rescuer, safety glass, gloves, and a warm jacket for exiting into cooler temperatures if present. Regardless of the level of experience of any tunneling practitioner, it is important that all visitors to a project site attend the safety orientation to be made aware of all site-specific work hazards and safety procedures that are often unique to any particular project site. In some cases, it may be necessary to attend safety orientations from both the tunnel constructor as well as the client.

Many underground works are geometrically complex and involve multiple locations of active construction. Tunnel practitioners should always include to bring water and a light snack during site visits during construction of a tunnel project in case of delays on egressing the underground works. For long tunnels, the worker train to the heading and

Figure 21.1 BE SAFE.

Figure 21.2 NO SAFETY – KNOW SAFETY.

return to the portal may be delayed due to the transport of materials necessary for tunnel construction and long waits can be expected.

Tunneling practitioners should only enter the underground works for visits and inspections when accompanied by the tunnel constructor in order that they are fully aware of the active works and any changes to safety procedures with the works. Tunneling practitioners should not enter underground works and headings where mucking is actively ongoing as part of drill and blast operations since it is usually dusty and visibility is typically poor which increases the risk of not being seen by the scooptram operator.

Tunneling practitioners should not enter the underground works and headings and perform an inspection and evaluation of the tunnel face when charging of the tunnel face is ongoing as part of drill and blast operations. This is to prevent any interference to the work crews and the introduction of unnecessary risks. Importantly, no sounding of rock exposures using a geological rock hammer should be performed during charging operations.

The stability of moderately to highly fractured rock conditions may in some cases be marginal even after the installation of rock support. Tunneling practitioners should prevent unnecessary scaling performed by the tunnel constructor in order to reduce the potential for the occurrence of instabilities.

All tunneling practitioners bear a responsibility to immediately stop all work and inform appropriate site staff of any safety risk that is identified during construction. Health and safety should be respected at all times during tunnel construction and it is a useful reminder to maintain multiple safety awareness signs throughout the project site to remind all personnel as presented in Figures 21.1 and 21.2.

References

AACE 18R-97. 2011. Recommended Practice: Cost Estimate Classification System as Applied in Engineering, Procurement, and Construction for the Process Industries. AACE International, 1265 Suncrest Towne Centre Dr., Morgantown, WV 26505 USA. Phone 800-858-COST/304-296-8444. Fax: 304-291-5728. Internet: http://www.aacei.org E-mail: info@aacei.org.

AACE 57R-09. 2011. Recommended Practice: Integrated Cost and Risk Analysis Using Monte Carlo Simulation of a CPM Model. Morgantown, WV: AACE International.

ASCE, 2007. Geotechnical Baseline Reports for Underground Construction, American Society of Civil Engineers. Ed. Essex, R.

Andreas Heier Sødal, A.H., Lædre, O., Svalestuen, F., and Lohne, J. 2014. Early Contractor Involvement: Advantages and Disadvantages for the Design Team. Proceedings from International Group for Lean Construction Conference, Oslo.

Agan, C. 2015. Prediction of Squeezing Potential of Rock Mass Around the Suruc Water Tunnel. Bulletin of Engineering Geology and the Environment 10064-015-0758-1. Springer Verlag Berlin.

Barla, G., and Barla, M. 2009. Innovative Tunneling Construction Method in Squeezing Rock, Kolloquium der Professur für Untertagbau der ETH Zürich, Tunnelbau in druckhaftem Gebirge, Zürich.

Barla, G., Barla, M., Bonini, M., and Debernardi, D. 2007. Lessons Learned During the Excavation of the Saint Martin La Porte Access Gallery Along the Lyon-Turin Base Tunnel. BBT 2007. Internationales Symposium Brenner Basistunnel uns Zulaufstrecken. Innsbruck University Press, pp. 45–52

Barneich, J. Majors, D., Moriwaki, Y., Kulharn, R., and Davidson, R. 1996. Application of Reliability Analysis in the Environmental Impact Report and Design of a Major Dam Project. ASCE.

Barton, N. September 1999. TBM Performance Estimation in Rock Using Qtbm. Tunnels and Tunnelling International, pp. 30–34.

Barton, N. Lien, R. and Lunde, R. 1974. Engineering Classification of Rock Masses for the Design of Tunnel Support. Rock Mechanics, Vol. 6, No. 4, pp. 189–229.

Benson, R.P. 1989. Design of Unlined and Lined Pressure Tunnels. Tunnelling and Underground Space Technology, Vol. 4, No. 2, pp. 155–170.

Best Practices for Roadway Tunnel Design, Construction, Maintenance, Inspection and Operations, 2011. United States National Cooperative Highway Research Program. Project 20-68A.

Bieniawski, Z. 1976. Rock Mass Classification in Rock Engineering. Exploration for Rock Engineering Symposium, Capetown.

Bieniawski, Z.T., Celada, B., Galera, J.M., and Alvarez, M. 2007. Rock Mass Excavability (RME) Indicator: New Way to Selecting the Optimum Tunnel Construction Method (Paper 06-0254). In *Proceedings of the ITA-AITES World Tunnel Congress and 32nd General Assembly*, Seoul, Korea.

Bieniawski, Z.T., Celada, B., Galera, J.M., and Rodrigues, A. 2012. Specific Energy of Excavation in Detecting Tunneling Conditions Ahead of TBMs. Tunnels and Tunneling International.

Bratveit, K., Bruland, A., and Brevik, O. 2016. Rock Falls in Selected Norwegian Hydropower Tunnels Subjected to Hydropeaking. Tunnelling and Underground Space Technology, Vol. 52, pp. 202–207.

Brekke, T.L., and Ripley, B.D. 1987. Design Guidelines for Pressure Tunnels and Shafts. Research Project 1745-17, AP-5273. Palo Alto, CA: Electric Power Research Institute.

Brierley, G.S., Corkum, D.H., and Hatem, D., eds. 2010. Design-Build: Subsurface Projects, 2nd ed. Littleton, CO: SME.

Broch, E. 2010. Tunnels and Underground Works for Hydropower – Lessons Learned in Home Country and Projects Worldwide. Sir Muir-Wood Lecture. World Tunnel Congress, Vancouver.

Brox, D.R. 2012. Evaluation of Overstressing of Deep, Hard Rock TBM Excavated Tunnels in British Columbia. Presented at the Tunnelling Association of Canada (TAC) Conference, Montreal, Quebec.

Brox, D.R. 2013a. Evaluation of Overstressing of Deep Hard Rock Tunnels, World Tunnel Congress 2013, Geneva, Switzerland.

Brox, D. 2013b. Technical Considerations for TBM Tunneling for Mining Projects. Society of Mining Engineers Transactions, Vol. 334, pp. 498–505.

Brox, D. 2016. Design and Functional Requirements of Rock Traps for Pressure Tunnels. The International Journal of Hydropower and Dams.

Brox, D.R., and Lee, K.W. 1995. Yielding and Collapse of Large Span Tunnels in Weak Rock, 8th Congress of International Society of Rock Mechanics, Tokyo, Japan.

Brox, D.R., and Hagedorn, H. 1998. Extreme Deformation and Damage during the Construction of Large Tunnels, IAEG 8th Congress / 15th Canadian Tunneling Conference, Vancouver, Canada. Also published in Tunneling and Underground Space Technology, 1999.

Brox, D.R., Moalli, S., Redmond, S., Procter, Jezek, D. 2008. TBM Tunneling at the Ashlu Creek Hydroelectric Project. 20th Tunneling of Association of Canada Conference, Niagara Falls, Ontario.

Bruland, A. 1998. Hard Rock Tunnel Boring. PhD thesis. Norwegian University of Science and Technology (NTNU) 1998:81.

Carter, T.G., Steels, D., Dhillon, H.S., and Brophy, D. 2005. Difficulties of Tunneling under High Cover in Mountainous Regions. In Tunnelling for a Sustainable Europe: AFTES [French Tunnelling Association and Underground Space] International Congress, Chambery, pp. 349–358.

Castro S.O., Van Sint Jan, M., Gonzalez, R.R., Lois, P.V., and Velasco, L.E. 2003. Dealing with Expansive Rocks in the Los Quilos and Chacabuquito Water Tunnels – Andes Mountains of Central Chile. ISRM 2003–Technology Roadmap for Rock Mechanics, South African Institute of Mining and Metallurgy.

Clark, J., and Chorley, S. 2014. The Greatest Challenges in TBM Tunneling: Experiences from the Field. North American Tunneling Conference, pp. 101–108.

Coastech Research, 2008. Acid Rock Drainage Prediction Manual: A Manual of Chemical Evaluation Procedures for the Prediction of Acid Generation from Mine Wastes. Mine Environment Neutral Drainage (MEND) Project 1.16.1b. CANMET No. 23440-9-9149/01-SQ SSC. CANMET-MSL Division, Department of Energy, Mines and Resources, Canada. Originally published 1991.

Concilia, M., and Grandori, R. 2004. The New Viola Water Transfer Tunnel. Mechanized Tunneling: Challenging Case Histories, International Congress Turin, GEAM.

Day, J. 2008. New Swiss Design Guidelines for Road Tunnels and How They are being Implemented, Second Brazilian Congress on Tunnels and Underground Structures, Sao Paulo.

De Biase, A., Grandori, R., Bertolo, P., and Scialpi. M. 2009. Gibe 2 Tunnel Project Ethiopia: 40 Bars of Mud Acting on the TBM, Special Designs and Measures Implemented to Face one of the Most Difficult Events in the History of Tunneling. In *Rapid Excavation and Tunneling Conference: 2009 Proceedings*. Edited by G. Almeraris and B. Mariucci. Littleton, CO: SME.

Deere, D., 2007. Tunneling through Mountain Faults. In *Rapid Excavation and Tunneling Conference: 2007 Proceedings*. Edited by M.T. Traylor and J.W. Townsend. Littleton, CO: SME.

Deere, D.U., and Deere, D.W. 1988. The RQD Index in Practice, Proc. Symp. Rock Class. Engineering Purposes, ASTM Special Technical Publications 984, Philadelphia, pp. 91–101.

Deere, D.W. 2007. Tunneling through Mountain Faults, Rapid Excavation and Tunneling Conference.

Design-Build – Subsurface Projects, 2002. Editors Brierley, G.S., and Hatem, D.J. Zeni House Publications.

Dickmann, T., and Krueger, D. 2014. How to Turn Geological Uncertainty into Mmanageable Risk? In *Proceedings of the World Tunnel Congress 2014 – Tunnels for a Better Life*. Foz do Iguaçu, Brazil.

Diederichs, M.S., Carter, T., and Martin, D. 2010. Practical Rock Spall Prediction in Tunnels. In *Proceedings from the International Tunnelling Association Conference*, Vancouver, Canada.

DIN 50929-3. 1985. *Corrosion of Metals*. Berlin: German Institute for Standardization.

Dutton, P., Stolz, J., Brady, J., and Ray, C. 2011. Design-Build Estimating from the Bottom Up. In *Rapid Excavation and Tunneling Conference: 2011 Proceedings*. Edited by S. Redmond and V. Romero. Englewood, CO: SME.

Edgerton, W., ed. 2008. Recommended Contract Practices for Underground Construction. Society of Mining Engineers.

Efron, N. and Read, M. 2012. Analysing International Tunnel Costs: An Interactive Qualifying Project, Worcester, MA: Worcester Polytechnic Institute.

Eskesen, S.D., Tengborg, P., Kampmann, J., and Veicherts, T.H. 2004. Guidelines for Tunnelling Risk Management: International Tunnelling Association, Working Group No. 2. Tunnelling Underground Space Technology, Vol. 19, No. 3, pp. 217–237.

Essex, R. 2008. Geotechnical Baseline Reports for Construction, Second Edition. In *North American Tunneling: 2008 Proceedings*. Edited by M.F. Roach, M.R. Kritzer, D. Ofiara, and B.F. Townsend. Littleton, CO: SME.

Federal Highway Administration, 2004. Road Tunnel Design Guidelines Manual. Us Department of Transportation.

Federal Highway Administration, 2005. Highway and Rail and Transit Tunnel Inspection Manual.

Federal Highway Department Administration, 2009. Technical Manual for the Design and Construction of Road Tunnels—Civil Elements. US Department of Transportation.

Frostberg, G., Carter, I., and Stenmark, M. 2007. Competitive Design Choices in an EPC Contract with an Extremely Compressed Time Schedule and High Liquidated Damages for Delays.

FHA and FTA (Federal Highway Administration and Federal Transit Administration), 2005. Highway and Rail and Transit Tunnel Inspection Manual. Washington, DC: U.S. Department of Transportation.

Fulcher, B., Bell, M., and Garshol, K. 2008. Pre-excavation Drilling and Grouting for Control of Water Ground Improvement in Highly Variable Ground Conditions at the Arrowhead Tunnel Project. In *North American Tunneling: 2008 Proceedings*. Edited by M.F. Roach, M.R. Kritzer, D. Ofiara, and B.F. Townsend. Littleton, CO: SME.

Galera, J.M., Paredes, M., Menchero, C., and Pozo, V. 2014. Risk Associated with Swelling Rocks in Volcanic Formations in the Design of Hydro-Tunnels. In *Rock Engineering and Rock Mechanics: International Society for Rock Mechanics* (ISRM) European Regional Symposium (Eurock), Vigo, Spain, May 26–28.

GEO (Geotechnical Engineering Office), 2009. *GEO Technical Guidance Note No. 25 (TGN25): Geotechnical Risk Management for Tunnel Works*, issue no. 2, revision A. Civil Engineering and Development Department, Government of the Hong Kong Special Administrative Region. Available from www.cedd.gov.hk/eng/publications/guidance_notes/doc/TGN25_2a.pdf.

Ghee, E.H., Zhu, B.T., and Wines, D.R. 2011. Numerical analysis of twin road tunnels using two- and three-dimensional modeling techniques. *Continuum and Distinct Element Numerical Modelling in Geo-Engineering*. Melbourne.

Goodfellow, R., and Mellors, T. 2007. Cracking the Code: Assessing Implementation in the United States of the Codes of Practice for Risk Management of Tunnel Works. In *Rapid Excavation and Tunneling Conference: 2007 Proceedings*. Edited by M.T. Traylor and J.W. Townsend. Littleton, CO: SME.

Grandori, R. 2011. Hard Rock Extreme Conditions in the First 10 km of TBM Driven Brenner Exploratory Tunnel. In *Rapid Excavation and Tunneling Conference: 2011 Proceedings*. Edited by M.T. Traylor and J.W. Townsend. Englewood, CO: SME.

Grandori, R. 2016. DSU TBM for Vishnugad Pipalkoti Project – TBM Design Development for Large Diameter Rock Tunnels Under the high Covers of the Himalaya, World Tunnel Congress, San Francisco.

Guidelines for Good Occupational Health and Safety Practice in Tunnel Construction, 2008. International Tunneling Association.

Guidelines for the Design of Tunnels, 1988. ITA Working Group on the Approaches to the Design of Tunnels. Tunnels and Underground Space Technology, Vol. 3, No. 3.

Guidelines for the Design of Shield Tunnel Lining, 2000. ITA Working Group No. 2. Tunnels and Underground Space Technology, Vol. 15, No. 3.

Hagedorn, H., Stadelmann, R., and Husen, S. 2008. Gotthard Base Tunnel Rock Burst Phenomena in a Fault Zone, Measuring and Modelling Results. Presented at the World Tunnel Congress, India.

Hedwig, P. 1987. A Theoretical Investigation into the Effects of Water Hammer Pressure Surges on Rock Stability of Unlined Tunnels. Underground Power Plants.

Hendron, A.J., Fernandez, G., Lenzini, P., and Hendron, M.A. 1987. Design of pressure tunnels In *The Art and Science of Geotechnical Engineering: At the Dawn of the 21st Century*. Edited by E.J. Cording. Englewood Cliffs, NJ: Prentice Hall.

Heslop, P., and Caruso, C. 2013. Recommendations on How Geotechnical Baseline Reports can be Prepared for Rock Tunnel Projects. In *Rapid Excavation and Tunneling Conference: 2013 Proceedings*. Edited by M.A. DiPonio and C. Dixon. Englewood, CO: SME.

Heuer, R. 1995. Estimating Rock Tunnel Water Inflow. RETC.

Heuer, R. 1995. Estimating Rock Tunnel Water Inflow. RETC.

Hoek, E. 1982. Geotechnical Considerations in Tunnel Design and Contract Preparation. Sir Julius Wernher Memorial Lecture, Tunnelling Symposium, Brighton, England.

Hoek, E. 2000. Big Tunnels in Bad Rock, Terzaghi Lecture. ASCE Journal of Geotechnical and Geoenvironmental Engineering, Vol. 127, No. 9, September 2001, pp. 726–740.

Hoek, E., and Brown. E.T. 1980. Underground Excavations in Rock. The Institute of Mining and Metallurgy, London.

Hoek, E., and Guevara, R. 2009. Overcoming Squeezing in the Yacambú-Quibor Tunnel, Venezuela. Rock Mechanics and Rock Engineering, Vol. 42, No. 2, pp. 389–418.

Hoek, E., and Marinos, P. 2000. Predicting Tunnel Squeezing Problems in Weak Heterogeneous Rock Masses. Tunnels and Tunnelling International, Part 1 – November 2000, Part 2.

Hoek, E., and Palmieri, A. 1998. Geotechnical Risks On Large Civil Engineering Projects. Keynote Address for Theme I – International Association of Engineering Geologists Congress, Vancouver, Canada.

Hoek, E. Carranza-Torres, C., and Corkum, B. 2002. Hoek-Brown Failure Criterion – 2002 Edition. In *Proceeding of NARMS-TAC Conference*, Toronto, Vol. 1, pp. 267–273.

Hoek, E., Carranza-Torres, C., Diederichs, M., and Corkum, B. 2008. The 2008 Kersten Lecture: Integration of Geotechnical and Structural Design in Tunneling. University of Minnesota 56th Annual Geotechnical Engineering Conference. Minneapolis.

Hoek, E., Wood, D., and Shah, S. 1992. A Modified Hoek-Brown Criterion for Jointed Rock Masses. In *Rock Characterization: International Society for Rock Mechanics Symposium (ISRM) European Regional Symposium* (Eurock). Edited by J.A. Hudson. London: British Geotechnical Association, pp. 209–214.

ICE (Institution of Civil Engineers), 2014. Tunnels: Inspection, Assessment and Maintenance. London: ICE.

International Society of Rock Mechanics Commission on Testing: 1999. Suggested Methods for Laboratory Testing of Swelling Rocks.

ITA (International Tunnelling and Underground Space Association) Working Group No. 3. 2011. Contractual Framework Checklist for Subsurface Construction Contacts.

International Tunnel Association (ITA) Working Group 5. Health and Safety in Works, 2008. Guidelines for Good Occupational Health and Safety Practice in Tunnel Construction.

ITA (International Tunnelling and Underground Space Association) Working Group No. 2. 2015. Strategy for Site Investigation for Tunnelling Projects.

Itasca Consulting Group, Inc. 2015. FLAC [software]. Minneapolis, MN. Available from www.itascacg.com.

Itasca Consulting Group, Inc. 2015a. FLAC [Version 8.0]. Minneapolis, MN. Available from www.itascacg.com.

Itasca Consulting Group, Inc. 2015b. UDEC [Version 6.0]. Minneapolis, MN. Available from www.itascacg.com.

ITIG (International Tunnelling Insurance Group). 2006. A Code of Practice for Risk Management of Tunnel Works. Available from www.imia.com/wp-content/uploads/2013/05/EP24-2006-Code-of-Practice-for-Risk-Management-of-Tunnelling-Works.pdf.

Kawata, T., Nakano, Y., Matsumoto, T., Mito, A., Pittard, F., and Honda, Y. 2013 Challenge in High-speed TBM Excavation of Long-Distance Water Transfer Tunnel, Pahang-Selangor Raw Water Transfer Tunnel, Malaysia. In *Proceedings of the World Tunnel Congress*, Geneva, Switzerland.

Kocbay, A., Marence, M., and Linortner, J. 2009. Hydropower Plant Ermenek/Turkey Pressure Tunnel – Design and Construction. In *Proceedings of the World Tunnel Congress*, Budapest, Hungary.

Kovari, K. 2012. Design Methods with Yielding Support in Squeezing and Swelling Rocks. World Tunnel Congress, Budapest.

Lang, T.A., Kandorski, F.S., and Kanwarjit, S.C. 1976. Effect of Rapid Water Pressure Fluctuations on Unlined Water Tunnel Stability. In *Rapid Excavation and Tunneling Conference: 1976 Proceedings*. New York: American Institute of Mining, Metallurgical, and Petroleum Engineers.

Lewis, W. 2009. Construccion de Tuneles Transandinos en la Cordillera de Los Andes. In *Proceedings of the Taller Internacional de Tuneles y Obras Subterranes*, Bogota, Columbia. (Spanish only)

Li, C., and Linbald, K. 1999. Corrosivity Classification of the Underground Environment. In *Rock Support and Reinforcement Practice in Mining*. Edited by A.G. Thompson, E. Villaescusa, and C.R. Windsor. Rotterdam, The Netherlands: A.A. Balkema.

MacKellar, D.C.R., and Reid, J.A.F. 1994. Mineralogical Aspects of the Durability of Basalts in the Lesotho Highlands. Presented at the South African Tunnelling Conference.

Martin, C.D., Giger, S., and Lanyon, G.W. 2016. Behaviour of Weak Shales in Underground Environments. Rock Mechanics and Rock Engineering, Vol. 49, No. 2, pp. 673–687.

Martin, P., and Farrukh, H. 2003. Electrical Resistivity: An Answer to the Challenges Raised by the Design of a 17 km-long Tunnel in the Andes. In *Rapid Excavation and Tunneling Conference: 2003 Proceedings*. Edited by R.A. Robinson and J.M. Marquardt. Littleton, CO: SME.

Merritt, A. 1999. Geological and Geotechnical Considerations for Pressure Tunnel Design. American Society for Civil Engineers (ASCE).

Mezgar, F., Anagnostou, G., and Ziegler, H.J. 2013. On Some Factors Affecting Squeezing Intensity in Tunnelling. Presented at the World Tunnel Congress, Geneva, Switzerland.

Microsoft Corporation, 2016. Microsoft Project Management software. Available from www. microsoft.com.

MinEx Health & Safety Council, 2010. *Industry Code of Practice: Underground Mining and Tunnelling.* Available from www.minex.org.nz/documents/COP%20Mining%20and%20Tunneling%20Feb10.pdf.

Montero, R., Victores, J.G., Martínez, S., Jardón, A., and Balaguer, C. 2015. Past, Present and Future of Robotic Tunnel Inspection. Automation in Construction, Elsevier.

Morald, J., and Kolenda, P. 2008. Design and Construction of Tunnel and Shaft Plugs. North American Tunneling.

New Austroad Guidelines for Tunnel Design, 2011. Australian Tunnelling Society.

NFPA 502. 2014. Standards for Road Tunnels, Bridges, and Other Limited Access Highways. Quincy, MA: National Fire Protection Association. Available from www.nfpa.org.

O'Carroll, J., Goodfellow, B., and Underground Construction Association of SME. 2014. Guidelines for Improved Risk Management on Tunnel and Underground Construction Projects in the United States of America. Englewood, CO: SME.

Okazaki, K., Mogi, T., Utsugi, M., Ito, Y., Kunishima, H., Yamazaki, T., Hashimoto, T., Ymamaya, Y., Ito, H., Kaieda, H., Tsukuda, K., Yuuki, Y., and Jomori, A. 2011. Airborne Electromagnetic and Magnetic Surveys for Long Tunnel Construction Design. Physics and Chemistry of the Earth, Parts A/B/C, Vol 36, No. 16, pp. 1237–1246, Urban Geophysics.

Oracle Corporation, 2016. Primavera P6 Professional Project Management software, Available from www.oracle.com.

O'Rourke. T. D., ed. 1984. Guidelines for Tunnel Lining Design, ASCE Geotechnical Engineering Division, Reston, Virginia.

Palisade Corporation, 2016. @RISK [software]. Ithaca, NY, Available from www.palisade.com

Palmstrom, A. 1995. RMi – A Rock Mass Characterization System for Rock Engineering Purposes. PhD thesis, Oslo University, Norway.

Panthi, K.K. 2006, Analysis of Engineering Geological Uncertainties Related to Tunnelling in Himalayan Rock Mass Conditions. PhD thesis. Department of Geology and Mineral Resources Engineering. Norwegian University of Science and Technology (NTNU), Norway.

Panthi, K.K., and Nilsen, B. 2007. Uncertainty Analysis of Tunnel Squeezing for Two Tunnel Cases from Nepal Himalaya. International Journal of Rock Mechanics and Mining Sciences, Vol. 44, No. 1, pp. 67–76.

Parker, H. 2004. 1 Congresso Brasileiro de Túneis e Estruturas Subterrâneas Seminário Internacional South American Tunnelling.

Perello, P., Venturini, G., Della Piane, L., and Martinotti, G. 2003. Geo-structural Mapping Applied to Underground Excavations: Updated Ideas after a Century since the First Trans-alpine Tunnels, Rapid Excavation and Tunneling Conference. In *Rapid Excavation and Tunneling Conference: 2003 Proceedings.* Edited by R.A. Robinson and J.M. Marquardt. Littleton, CO: SME.

Piaggio, G. 2015. Swelling Rocks Characterization: Lessons from the Andean Region. World Tunnel Congress, Dubrovnik.

Petrofsky, A.M. 1987. The Jacobs Sliding Floor: Current Competitive Applications. VI Australian Tunneling Conference, Melbourne.

Plinninger, R., Kasling, H., Thuro, K., and Spaun, G. 2003. Testing Conditions and Geomechanical Properties Influencing the CAI Value. International Journal of Rock Mechanics and Mining Sciences, Vol. 40(2003), pp. 259–263. Permagon Press.

Price, W.A., and Errington, J.C. 1998. Guidelines for Metal Leaching and Acid Rock Drainage at Minesites in British Columbia. Fort St. John, BC: Ministry of Energy and Mines.

Ramoni, M. 2010. Feasibility of TBM Drives in Squeezing and the Risk of Shield Jamming, ETH Dissertation 18965, Swiss Federal Institute of Technology, Zurich.

Rancourt, A., Design of Unlined Pressure Tunnels, International Tunneling Association World Tunnel Congress 2010, Vancouver, Canada.

Reilly, J., and Brown, J. 2004. Management and Control of Cost and Risk for Tunneling and Infrastructure Projects. In *Underground Space for Sustainable Urban Development: Proceedings of the 30th ITA-AITES World Tunnel Congress.* Singapore.

Reilly, J., Sangrey, D., and Warhoe, S. 2010 Management of Cost and Risk to Meet Budget and Schedule, North American Tunneling.

Richard, J.A. 1998. Inspection, Maintenance and Repair of Tunnels. Tunneling and Underground Space Technology, Vol. 13. No. 4, pp. 369–375.

Rocscience. 2015a. Dips [Version 7.0]. Toronto, ON. Available from www.rocscience.com.

Rocscience. 2015b. Phase2 [Version 9.0]. Toronto, ON. Available from www.rocscience.com.

Rocscience. 2015c. Unwedge [Version 4.0]. Toronto, ON. Available from www.rocscience.com.

Rocscience. 2016. RocSupport [Version 4.0]. Toronto, ON. Available from www.rocscience.com.

Road Safety in Tunnels, 1996, World Road Association.

Road Tunnel Safety Regulations, 2007. United Kingdom Ministry of Transport.

Rønning, J.S., Ganerod, G.V., Dalsegg, E., and Reiser, F. 2014. Resistivity Mapping as a Tool for Identification and Characterisation of Weakness Zones in Crystalline Bedrock: Definition and Testing of an Interpretational Model. Bulletin of Engineering Geology and the Environment, Vol. 73, No. 4, pp. 1225–1244.

Rosin, S. 2005. Geotechnical Risk Assessment and Management for Maintenance of Water Conveyance Tunnels in Southeastern Australia. Presented at the Australian Geotechnical Society–Australian Underground Construction and Tunnelling Association Mini-Symposium: Geotechnical Aspects of Tunnelling for Infrastructure Projects.

Rothfuss, B., Bednarz, S., and Clarke, E. 2011. Water Tunnel Condition Assessment: A Comprehensive Approach to Evaluating Reliability. Hydrovision.

Safe Working in Tunneling for Tunnel Workers and First Line Supervision. 2011. International Tunneling Association.

Salter, W.O. 1992. ITA Recommendations n Contractual Sharing of Risks: ITA Working Group on Contractual Sharing of Risks. Tunnelling Underground Space Technology, Vol. 7, No. 4, pp. 393–397.

Saw, H.A., Villaescusa, E., and Windsor, C. 2015. Safe Re-entry Time with In-Cycle Shotcrete for Support of Underground Excavations. Presented at the Shotcrete for Underground Support Conference XII. Singapore.

Schmach, P., and Peters, M. 2016. Mechanized Tunneling Solutions for Small Hydro Plants. International Journal on Hydropower and Dams, Vol. 23, No. 1.

Schwartz, E. 2004. Austrian Guideline for Geomechanical Design of Tunnels—Necessity for Cooperation between Geologists, Geotechnical and Civil Engineers. HYPERLINK "http://link.springer.com/book/10.1007/b93922" Engineering Geology for Infrastructure Planning in Europe. Volume 104 of the series Lecture Notes in Earth Sciences pp. 39–46.

Sepehrmanesh, M., Rostmai, J., and Gharahbagh, E. 2012. Planning Level Tunnel Cost Estimation Based on Statistical Analysis of Historical Data. In *North American Tunneling: 2012 Proceedings.* Edited by M. Fowler, R. Palermo, R. Pintabona, and M. Smithson Jr. Englewood, CO: SME.

SIA 197. 2004. Design of Tunnels: Basic Principles. Zurich, Switzerland: Swiss Society of Engineers and Architects. Available from www.sia.ch/en/the-sia/the-sia/.

SIA 198. 2004. Underground Structures: Execution. Zurich, Switzerland: Swiss Society of Engineers and Architects. Available from www.sia.ch/en/the-sia/the-sia/.

Steiner, W., Kaiser, P.K., and Spaun, G. 2010. Role of Brittle Fracture on Swelling Behaviour of Weak Rock Tunnels: Hypothesis and Qualitative Evidence. Geomechanics and Tunnelling, Vol. 3, No. 5, pp. 583–596.

Stolz, J. 2010. Overhead and Uncertainty in Cost Estimates: A Guide to their Review. In *North American Tunneling: 2010 Proceedings*. Edited by L.R. Eckert, M.E. Fowler, M.F. Smithson Jr., and B.F. Townsend. Littleton, CO: SME.

Sturk, R., Dudouit, F., Aurell, O., and Eriksson, S. 2011. Summary of the First TBM Drive at the Hallandas Project, Rapid Excavation and Tunneling Conference. San Francisco, USA.

Terron, J. 2014. Manual of Construction of Tunnels through Faults (in Spanish).

Terzaghi, K. 1946. Rock Defects and Loads in Tunnel Supports. In *Rock Tunneling with Steel Supports*. Edited by R.V. Proctor and T.L. White. The Commercial Shearing and Stamping Co., Youngstown, Ohio, pp. 17–99.

Thapa, B., Nitschke, A., Ramirez, I., McRae, M., and Vojtech, G. 2013. Lessons Learned from NATM Design and Construction of the Caldecott Fourth Bore, RETC Washington, DC.

Thompson, A. 2013. GBR—TO USE OR NOT TO USE? Rapid Excavation and Tunneling Conference.

Türtscher, M. 2016. Simtunnel Pro 2.0 [software]. Available from www.simtunnel.com.

Tunnels: Inspection, assessment and maintenance, 2014. Institute of Civil Engineers, London.

Tunnel Lining Design Guide, 2004. British Tunneling Society. Institute of Civil Engineers, London.

Urschitz, G., and Gildner, J. 2004. SEM/NATM Design and Contracting Strategies North American Tunneling Conference, Atlanta, USA.

US Army Corps of Engineers 1997. Tunnels and Shafts in Rock. EM 1110-2-2901.

Wahlstrom, E.E. 1973. Tunneling in Rock. Developments in Geotechnical Engineering 3, Elsevier.

Wang, J. 1993. Seismic Design of Tunnels, A Simple State-of-the-Art Design Approach. 1991 William Barclay Parsons Fellowship, Parsons Brinckerhoff, Monograph 7.

Werner, D., and Wannemacher, H. 2009. Alborz Service Tunnel in Iran: TBM Tunneling in Difficult Ground Conditions and its Solutions. Presented at the 8th Iranian Tunneling Conference, Tehran.

Wirthlin, A., Fusee, R., Nolting, R., Sun, Y., and Tsztoo, D. 2016. Squeezing Ground: Conditions & Lessons Learned at the New Irvington Tunnel, World Tunnel Congress, San Francisco.

WorkCover. 2006. Tunnels under Construction: Code of Practice. New South Wales, Australia: WorkCover. Available from www.workcover.nsw.gov.au/__data/assets/pdf_file/0018/20772/Tunnels-Under-Construction-Code-of-Practice.pdf.

Yagiz, S., Rostami, J., and Ozdemir, L. 2012. Colorado School of Mines Approaches for Predicting TBM Performance. In *Rock Engineering and Technology for Sustainable Underground Construction: International Society for Rock Mechanics (ISRM) Symposium European Regional Symposium* (Eurock) Conference, Stockholm, Sweden.

Zangerl, C., Evans, K., Eberhardt, E., and Loew, S. 2008. Consolidation Settlements above Deep Tunnels in Fractured Crystalline Rock. Part 1 – Investigations above Gotthard Highway Tunnel. International Journal of Rock Mechanics and Mining Sciences, Vol. 45, No. 8, pp. 1211–1225.

Zhang, J., and Morgan, D.R. 2015. Quality Control for Wet-Mix Fiber Reinforced Shotcrete in Ground Support. Presented at the Shotcrete for Underground Support Conference XII. Singapore.